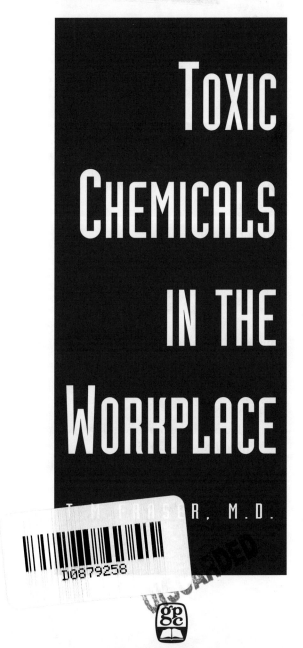

Toxic

Chemicals

in the

Workplace

T. M. FRASER, M.D.

gpgc

Gulf Publishing Company
Houston, London, Paris, Zurich, Tokyo

Toxic Chemicals in the Workplace

Gulf Publishing Company
Book Division
P.O. Box 2608 ☐ Houston, Texas 77252-2608

10   9   8   7   6   5   4   3   2   1

Library of Congress Cataloging-in-Publication Data
Fraser, T. M. (Thomas Morris), 1992–
    Toxic chemicals in the workplace /
T. M. Fraser.
        p.   cm.
    Includes bibliographical references and index.
    ISBN 0-88415-871-3
    1. Industrial toxicology.   2. Industrial hygiene.   I. Title.
RA1229.F73   1996
615.9′02—dc20   95-42879
                                                      CIP

Contents

Section Two: Five Steps to a Healthier Workplace

CHAPTER NINETEEN

Step Five: Control the Exposure—Personal Protective Equipment

APPENDIX ONE

Hazards from Various Processes

APPENDIX TWO

Short-term Detector Tubes

APPENDIX THREE

Useful Sources

Preface

In recent years, questions of health and the environment have become of paramount concern to politicians and the public at large. In particular, the state of health in the workplace has, of necessity, absorbed the interest of various government departments, public institutions, labor unions, and industrial organizations. As a result, many managers now find themselves in a position in which the law, the demands of labor, and their own humanitarian principles require that they take some kind of action to improve the health of their workplace. And while that is a desirable objective, all too often some of these managers may not know what action ought to be taken. This book addresses that problem. It is concerned with the recognition, evaluation, and control of toxic chemicals in the workplace, and in particular, with the management of some common workplace chemicals.

It is true that various textbooks, reference books, and sundry documents discuss this subject matter. This book, however, takes a more practical approach to the management of toxic chemicals in the workplace than is normally found in the more academic textbooks. In particular, it avoids jargon and technical terminology, yet it is sufficiently comprehensive to provide authoritative practical knowledge without going into esoteric theory. While the book is oriented particularly to the manager who has a university or community college background, it is also applicable to those persons who have technical interests and responsibilities but may have only a modicum of specific technical knowledge. Consequently, although dealing with what has been defined as industrial toxicology and some aspects of human pathology and industrial engineering, the content is presented in a manner that's easy to understand. All technical and otherwise unfamiliar concepts are explained in nontechnical language.

I developed the idea of the book as a result of requests from various participants at seminars I conducted. These participants included production managers, human resources managers, safety and health managers, industrial engineers, government labor inspectors, workers' compensation personnel, labor union health representatives, and professionals in occupational health and occupational hygiene, as well as students in these areas. These groups and others who need authoritative information on the topics find existing texts too academic, too research oriented, and too little directed to the practical requirements of the shop floor.

This book is divided into two sections. The first section provides the background to allow readers a knowledgeable understanding of the subject so they can recognize and evaluate whatever potential or actual problems may beset their workplace. The second section presents five simple steps readers can use to deal with the problems that have been defined.

A unique feature of the book is the inclusion of case studies, or "cautionary tales," which are fictionalized versions of real-life occurrences that are interspersed between chapters of factual content. These fictionalized tales are based on my experience and reflect some of the more interesting and often more dramatic aspects of chemical problems in the workplace.

So, go ahead. As the great 16th-century English essayist Francis Bacon once wrote: "Read, mark, learn, and inwardly digest." I only hope the material is not too indigestible.

T. M. Fraser

Recognition and Evaluation

CHAPTER ONE

What's the Problem?

People have been working since time immemorial. They have worked to survive; they have worked for profit, and they have worked for pleasure. They have worked from dawn to dusk and from dusk to dawn. Unfortunately, for hundreds of years, many workers have paid the price for their diligence in crippled limbs, disabled bodies, sickness, and death.

The great Italian physician Ramazzini,* who is considered to be the father of occupational medicine, wrote in the early 1700s about the health problems that beset the tradesmen of his time—the cobblers, the coopers, the tanners, the millers, and the others—who labored from childhood to middle age in work that they knew would ultimately kill or disable them. He described trades in which children entered the workforce at the age of 10 or 12 even though they knew that by the age of 40 they could be disabled or even dead. Although Ramazzini described the conditions and treated the sick, he couldn't do much to prevent these conditions. It is only now, in fact, over the past 40 years or so, that society has begun to recognize that a healthy workplace is not merely a desirable luxury, but a right for the worker and an asset to productivity.

All too often, however, the persons who are responsible for providing and maintaining a healthy workplace are not too sure about how to go

---

*Ramazzini, B., *De Morbis Artificum* (1713). Chicago: University of Chicago Press, translated by W. C. Wright, 1940.

about it and are a little wary of dabbling in a field that has largely been left in the hands of specialists. In fact, a curious paradox can be found in some industrial plants where managers receive no training in occupational health and hygiene and are less knowledgeable in those areas than some of their employees who attend special courses on health and hygiene given by various worker organizations. So what's the nature of the problem?

Most factory workers spend at least eight hours out of the 24, and more than half of their lifetime, on the shop floor. Most workers are probably unaware that over 100,000 toxic or potentially toxic chemicals, any of which might be found in some workplace, are listed in the *Registry of Toxic Effects of Chemical Substances* published by the National Institute for Occupational Safety and Health of the U.S. Department of Health, Education, and Welfare. Thousands of chemicals are added to the list every year. A few of these are thoroughly investigated to determine their toxic potential before they are introduced to the workplace. Some are classified as toxic on the basis of their chemical structure, or from experience, or perhaps because of the known toxicity of some related chemical, but in fact, relatively little is known about the toxic effects of many. Often, indeed, it is only after someone experiences an adverse effect from a given chemical that a thorough investigation of its potential toxicity is made. The long-term effects may not even be suspected. Furthermore, it is only in relatively recent years that anyone considered the possibilities of fetal damage from chemicals in an apparently healthy mother. And all too often, the worker on the shop floor, or even the manager, has very little knowledge of what's toxic and what's not. Thus, for a greater or lesser part of his or her working life, a worker, depending on the nature and conditions of the work, can be unwittingly exposed to potentially harmful chemicals.

Health, is course, is more than just the absence of sickness, and there are many aspects of health in the workplace—social, psychological, and physiological—that we are not concerned with here. What we are concerned with are the problems that derive from exposure, potential and actual, to toxic materials placed wittingly or unwittingly in the working environment. But let's go back a little to see how management of health in the workplace began.

HISTORY OF HEALTH MANAGEMENT

For centuries before the Industrial Revolution, other than for craft workers in certain trades such as mining, metal working, cobbling, tanning, felt making, and plumbing, most persons had little contact with hazardous chemicals. Any control of worker exposure that was deemed necessary (and that was very little) was exerted by the craft guilds or by common law relating to public nuisance. However, with the Industrial Revolution, which began in Britain in the 1700s, things began to change. The Industrial Revolution in Britain brought thousands of uneducated, untrained, totally impoverished, and relatively ignorant workers into the burgeoning workplace. Exploitation was the norm. Child labor and women's labor, under intolerable conditions, were taken for granted. Illness and impairment were rife, and death was not uncommon. It became obvious to thinking persons that some kind of reform was essential.

The big boost for reform came, in fact, with the Factory Acts in England. Three such acts were passed between 1833 and 1867, and they did much to improve working conditions and, in particular, initiated the training and use of "Factory Inspectors" to ensure that the provisions of the acts were implemented. These factory inspectors grew into the labor inspectors, safety officers, and occupational hygienists of today.

More was yet to come, however. Under pressure from an increasingly informed public, as well as from the legal profession and the trade unions, and following the social legislation and occupational accident insurance introduced by Bismarck in Germany in the 1880s, a significant development occurred in 1897. The promulgation of the Workmen's Compensation Act in Britain in that year provided for compensation for an injured worker except in the case of willful misconduct. That compensation, which is similar to that provided for today in the various states in the U.S. and elsewhere, was financed by a levy on employers and was based on proportional payments arising from the intrinsic hazards in the industry, as modified for any given company by its specific health and injury record within that industry. The original act was limited in scope to a few industries and situations, but it was expanded in 1907 and ultimately became a world model that included occupational illness as well as injury. New Jersey and Wisconsin were the first American states to follow suit in 1911, and Ontario became the first in Canada in 1915. The concept spread throughout Europe, the Americas, Australia, and other English-speaking countries, until workers' compensation or state- and company-financed insurance became established in developed countries around the world.

These reforms produced significant effects, not the least of which was a general improvement in working conditions throughout major industries. Much remained to be done, however, particularly during and following the worldwide depression of the thirties and the disruptions of World Wars I and II, all of which slowed the potential for any rapid improvement. Indeed, as industry became more complex, and as the use of sophisticated chemicals became more widespread, newer and more serious problems began to beset the workplace. Governments throughout the world again found themselves under pressure to take control and ensure the safety of the workforce. Institutions such as the International Labour Office (ILO), as well as the National Institute for Occupational Safety and Health (NIOSH), and the Occupational Health and Safety Administration (OSHA) in the United States provided guidance and ultimately legislation. In 1970, the United States passed the Occupational Safety and Health Act and in 1976, the Toxic Substances Control Act. Other state and provincial acts began to be promulgated in the U.S., Canada, Europe, and elsewhere.

The net result, unfortunately, was that, while conditions were improved to some extent, a surfeit of laws, regulations, and guidelines appeared to control and preserve health in the workplace. Each law or guideline required attention and action, while all too often, management and the worker on the shop floor found themselves with little real understanding of the nature or even existence of their problems. They might realize they had problems and that the law demanded action. For that matter, their humanitarian instincts might tell them that conditions could be improved, or they might even suspect that if conditions were improved, production and ultimately profit would also be improved, but because they didn't really know what the problems were, they didn't really know how to go about fixing them. Often, even if they could identify the problems, they still didn't know how to go about fixing them in a systematic and cost-effective way.

The question of cost effectiveness raises still another problem. Occupational health statistics are hard to come by. It is relatively easy to get statistics on accident absenteeism or even on sickness absenteeism occasioned by some sudden and unexpected exposure to, for example, a toxic gas. But sickness absenteeism arising from some prolonged and unrecognized low-level exposure may not even be recognized as such, and certainly never gets recorded as an occupational illness. Few plants, indeed, have the medical or even nursing services to maintain comprehensive records of sickness absenteeism from which such illness might be recognized. Most sickness is dealt with by outside physicians who provide minimal information to the patient's workplace. Thus, while it is relatively easy to show a cause-and-effect relationship between unsafe conditions and injury, it is often difficult to show a cause-and-effect relationship between unhealthy conditions and illness, even if one can demonstrate the presence of the unhealthy conditions.

At the same time, it is intuitively obvious that a plant in which working conditions are conducive to good health is likely to be more productive than one in which they are not. Absenteeism in such a plant is lower, work satisfaction is greater, and morale is higher. Some of the major world corporations, such as Dupont and General Motors in the U.S. and Volvo in Sweden, who have the capacity to conduct the necessary studies, have demonstrated with fact and figure that where working conditions are good, production is higher.

So in principle, at least, when we consider health in the workplace, we're not dealing with anything that hasn't been thought about before. Much has been done to improve overall health conditions, but much still remains to be done in the individual workplace. The gross dangers and inequities that roused the endeavors of the reformers a century and a half ago have disappeared. The days of cynical exploitation of men, women, and children in an inhumane environment are all but gone. Policy makers, legislators, and specialists of one sort or another have had their say. Laws have been passed, regulations promulgated, and guidelines provided. In fact, we are sometimes overwhelmed with demands and information, much of which may have little bearing on the problems at hand in our own particular bailiwicks. Sometimes we may wonder whether we are doing the right thing, whether we are doing enough, or whether we are doing too much. Is our work environment as healthy as it reasonably can be? Can it be improved, and if so, how can we go about improving it? Should we concentrate on the problems of the one or two new chemicals that enter our own specific working environment, or should we, in fact, reconsider the totality of that environment? Have we become too complacent about the ordinary industrial chemicals with which we may come into contact with each day—the heavy metals, the solvents, the dusts, and the industrial gases? These are some of the questions that have to be considered, and, it is hoped, will be answered here.

FINDING SOLUTIONS

Chemicals, unfortunately, don't sort themselves out into neat groups that behave the same way in the human body. Just because one chemical has much the same structure as another, it doesn't necessarily behave the same way as the other. What we want to know is which ones are toxic, that is, capable of causing bodily harm and how much harm they can cause. Unfortunately, we can't always do that just by looking at the chemical structure. Carbon *monoxide,* for example, which is a chemical combination of the soot called carbon and the gas called oxygen, is a poisonous gas which, when inhaled into the body, limits the amount of life-giving oxygen that can be carried by the blood. Carbon *dioxide,* on the other hand, which is also a combination of carbon and oxygen, is not poisonous at all. It is produced in the body in the course of ordinary living, and large quantities are breathed out from the lungs. So we can't classify toxic materials merely by their structure. Nor can we classify them in groups solely according to their function. One solvent, for example, such as carbon tetrachloride has primary effects on the brain, liver and kidney; another such as MEK (methyl ethyl ketone) primarily affects the breathing passages, eyes, and skin.

Clearly it is not feasible to describe all the chemicals that one could come across in all different types of work, nor would it be all that useful. In point of fact, in most jurisdictions it is mandatory that the manufacturer of a chemical used in the workplace must provide information to the user on the nature of that particular chemical, its potential hazard, and the steps that should be taken both to protect against it and to treat any emergency exposure that may occur. This information document is generally known as an MSDS (Manufacturer's Safety Data Sheet) and should be available to all concerned in any place where that chemical is in use. This book is not intended to usurp the MSDS. In all cases in which a strange chemical is in use, the MSDS should be consulted. This book is intended to acquaint the reader with some of the problems derived from ordinary, everyday chemicals.

So what this book proposes—and this will probably upset both the chemical purists and the toxicologists—is to look at toxic materials partly in terms of their effects and partly in terms of their chemical structure. Using several examples and case studies, we'll look at the overall nature of chemical toxicity in the workplace, with particular emphasis on some of the more commonplace chemicals in the form of:

- irritant gases and vapors
- choking gases
- solvents
- heavy metal poisons
- metal fume fever producers
- chemicals that damage the blood cells
- chemicals that cause sensitivity

But before we begin that discussion, let us look at a strange occurrence that happened in a certain factory where, even if you looked quite closely, the working conditions seemed to be ideal.

And yet . . .

CHAPTER TWO

Mrs. Madison's Mysterious Malady

Progessive Telesystems, as its name would imply, is a very progressive company. As a subsidiary of a major telecommunications corporation, its major products are telephone switchboards and their associated compo nents. During the 25 years of its existence, the company expanded considerably until it could no longer be contained within the original warehouse-type building in which it started. Consequently, it built an annex at the rear of the building, where 25 ladies could sit at workstations assembling circuit boards. The annex is rectangular and juts out of the back of the main building like the leg of a T. Inside, the workstations are organized in two rows, each with 12 stations, with one workstation between the two rows at the far end. The work is a little tedious and finicky, but not hard. The stools on which the ladies sit are comfortable and adjustable. The benches and frames where they work are well lit, with tools, parts, and components so organized as to be readily available as they are required. Each workstation has an adjustable footstool. There is even carpet on the floor and pleasant music playing during the working day. The annex is heated in winter to a comfortable temperature and cooled in the summer. Although it is not considered to be a "clean room" of the type in which some high precision electronic equipment is assembled, it is to some extent isolated. It has its own ventilation system with special air intakes on the roof of the building and outlets into the room through the ceiling around the periphery. The doors and windows are normally kept closed. The windows, in fact, cannot be opened.

Virtually no chemicals are used in the assembly. Wires and components are fastened in place by ingenious clips, with almost no soldering. Occasionally, but certainly not commonly, some small quantities of solvent are used for cleaning special applications.

Working conditions were about as favorable as could be when Mrs. Madison started working there. Indeed, the employees vied with each other to get transferred to the annex, although for various reasons—and this is not the place to go into politics—only 25 of the most senior ladies were selected. And then, three weeks after the annex was opened, the trouble started on a cold blustery day in November.

The wind was blowing hard when Mrs. Madison got to work that morning, although it only lasted an hour or so. Mrs. Madison is a 45-year-old woman who has been with the company for about 15 years. She is a good worker and tends to be a leader in the group, although she has never sought supervisory status. In fact, she sits by herself at the somewhat isolated work station at the far end of the annex, right under one of the air outlets, a spot she chose because she liked the flow of the cool air in summer and the warm air in winter. Mrs. Madison has never been a health worrier, nor has she been in the habit of complaining about minor discomforts. But when she began to develop a blinding headache about an hour after she got to work that morning, she was somewhat concerned. The headache did not go away even after she took a couple of pain pills. In fact, it got worse. And so, when another ten minutes passed without relief, she excused herself to her supervisor and went to see her doctor. That, unfortunately, wasn't much help because he couldn't find anything wrong with her, but he sent her home to recover. The headache lasted all day, but by the following morning, it was gone and she went back to work.

Nothing much happened for another couple of weeks, and then on another windy day, with a hint of snow, the headache struck again, fairly rapid in onset, blinding, disabling. Hoping it would go away, she stuck it out this time. But when she mentioned it to her two workmates some 10 feet on each side of her, they exclaimed, almost in unison: "That's strange! I've got a headache too!"

Then, from all over the room, the ladies began to complain in varying degrees of intensity that they too had headaches.

At this point, Mrs. Madison stood up. As she stood up, she felt dizzy and had to hang on to the bench. She stood still for about 10 seconds, feeling weaker and weaker, until she fell to the floor in a faint. Mrs. Jensen and Mrs. Kvalski stood up to help and fell in a faint beside her. Suddenly, all over the room, people began to stand up and promptly sit or fall down in varying degrees of faintness.

When the ambulances arrived ten minutes later, they took 12 of the 25 women to the hospital. They all went home the following day, none the worse. And no one at the hospital could say what the problem was.

That's a true story—more or less. As they say on TV, the names have been changed to protect the innocent. So what happened? Why did all the women get headaches? Why did Mrs. Madison get sick first? Why did she and some of the others get dizzy and faint? The immediate assumption, of course, was that something in the workplace had caused the problem. But what? There was nothing harmful in the work itself. Virtually no chemicals were used, except a little cleaning solvent on occasion. No other materials of significance were around, and the annex itself was an almost isolated and self-contained room.

This problem preplexed the hygienists sent in from the parent corporation and the government health inspectors and the consultant hired by the company.

The panic settled down, however. The ladies went back to work. Mass hysteria, the pundits said. A bunch of women reacting blindly because their leader got upset by something.

And then about three weeks later, another windy day as it turned out, the mysterious malady struck again. Mrs. Madison felt the headache coming on and knew what was going to happen. But this time there was no panic. She and the supervisor got the rest of the women together, according to a previously established procedure, and went back into the main building. And this time, a fortunate circumstance occurred. One of the hygienists was still in the building doing some follow-up air sampling. He had a bit of a headache himself, but as he was about to leave with the others, he noticed a curious phenomenon. The cheeks of the normally sallow Mrs. Madison showed a bright flush, not the red of overexertion or excitement, but a curious pinkish tinge—a symptom, which to his trained eyes, and combined with the headache, was almost diagnostic of carbon monoxide poisoning. As quickly as he could, he grabbed his test kit. Sure enough, carbon monoxide was present in the air in excessive quantities. OK! So that solved one problem. But where did it come from?

Back came the government inspectors. Back came the consultant. The hygienists tested every inch of the room for sources—the workstations, the floor, the ceiling, the doors, the windows, the ventilator inlets, including the one over Mrs. Madison's workstation. But nowhere could they find any carbon monoxide. In fact, they began to doubt the veracity of the hygienist who made the original discovery, even when he produced the tell-tale testing tube from his kit.

Then one day, Mrs. Madison had an idea. Suppose, she thought, the carbon monoxide is only there sometimes—perhaps only on a windy day, perhaps indeed only when the wind comes from a certain direction. She mentioned the idea to the boss. He looked a little skeptical but thought he'd better follow through, and so, with the maintenance engineer, he climbed onto the roof of the Annex—a flat roof spouting pipes and funnels, ventilator intakes, and exhaust gas outlets. And sure enough, there was the answer. The opening to the Annex ventilation system pointed directly towards the opening of the exhaust gas outlet from an oven in another part of the building, with only five feet separating the two. Under normal circumstances, when the prevailing wind blew the exhaust gas away from the ventilation inlet, no carbon monoxide entered the system. On the rather rare occasions when a strong wind blew towards the ventilation system, a continuous blast of carbon monoxide poured in, straight to Mrs. Madison's workstation and eventually to the rest of the room. Because she was the first and most direct recipient, she got the worst of the blast, although the others weren't too far behind.

Carbon monoxide is a clear and colorless gas, with no odor. Mrs. Madison was aware of nothing—only the blinding headache and the weakness that finally incapacitated her and her friends.

So what's the moral of this particular tale? Well, first, the story shows that exposure to a dangerous material, in this case carbon monoxide, can occur in the most unlikely situation. Second, the problem can be identified with diligent work (and sometimes a little luck!); and third, sometimes some unforeseen phenomenon, as in this case an improperly oriented exhaust gas outlet and an unexpected wind, can cause a problem that otherwise wouldn't exist. And, of course, the story also shows that sometimes you can't recognize a need until it is thrust upon you.

So with that behind us, let's take a look at some of the things we need to know before we go any further, such as, how much do we need to know about chemistry, what is meant by toxicity, when does a toxic chemical become a hazard, and how much is too much? We'll look at these things in the next chapters.

CHAPTER THREE

Chemistry and the Workplace

WHAT IS CHEMISTRY AND WHAT DO WE NEED TO KNOW?

Back in the 16th century, the great alchemist, metallurgist, and perhaps charlatan known as Paracelsus spent much of his energies, when he was not wenching or drinking, trying to turn lead into gold. In so doing, he became for a while the head of the medical school at one of the great medieval universities, as well as an expert on mercury and many other poisons. Many a medical school head has since tried to turn the equivalent of lead into gold, with an equal lack of success. However, in the course of an extraordinary body of work, some of it practical, some of it mystic, Paracelsus developed the science of chemistry. Chemistry wasn't new in his time, but what there was of it existed as haphazard mishmash of philosophy, magic, and chemical dabbling that went by the name of alchemy. What Paracelsus did, amongst many other achievements, was to begin the systematic organization of chemistry as we know it today and to develop, in primitive form, many of the techniques that are still used in principle for analyzing and processing chemicals.

So what is chemistry? Obviously, chemistry is the study of chemicals or materials. It turns out, however, that there are two basic types of chemistry, one known as *inorganic chemistry* and the other as *organic chemistry*.

INORGANIC CHEMISTRY

Inorganic chemistry is the study of the structure, relationships, and interactions of all the nonliving materials that make up the earth's crust, as well as the waters and the atmosphere. Thus, it includes the study of metals, such as iron, mercury, and lead; of gases, such as oxygen, hydrogen, and sulphur dioxide; of acids, such as sulphuric and hydrochloric; of salts, such as sodium chloride (common salt) and potassium chloride, and so on—all the chemicals, in fact, that are not, or have not been, part of living tissue.

Many inorganic chemicals under the proper circumstances can combine with other inorganic chemicals. Thus, for example, iron can be oxidized, or combined with oxygen, to form iron oxide, or, as we commonly know it, rust. The yellow powdery element called sulfur can be oxidized to form the dangerous and toxic gas sulfur dioxide, which in turn can unite with water to form the equally dangerous sulfuric acid. The metal sodium, which has the curious property of burning and exploding when placed on water, can combine with the potent war gas, chlorine, to form harmless sodium chloride, or common salt, which is essential for life. Nitrogen, which is the most common element in the air that we breathe, can, under certain circumstances such as welding, combine with life-giving oxygen to form a number of different, but deadly poisonous "nitrous fumes," which can harm or even kill the unwary. Thus many harmless or relatively harmless chemicals, when combined with other harmless chemicals, can become hazardous.

Some of these hazards will be considered at greater length later in this book. In the meanwhile, what about organic chemistry?

ORGANIC CHEMISTRY

Organic chemistry is the study of those chemicals which are, or have once been, part of living materials. Fundamentally, it is the study of the relationships and interactions of the chemical known as carbon, which in turn is the element that forms the basic structure of all living things.

Carbon is an element found in its simplest form as charcoal or the black material of soot or ink. Curiously, carbon can exist in another physical form that we call diamond. But that's irrelevant at this point.

As already noted, it is the basic element of all living and once-living tissue. It has the capacity to combine in a multiplicity of complex chains with other elements to form new chemicals. Some of these you know as carbohydrates, fats, and proteins. But there are many others. There are many different carbohydrates, fats, and proteins that in turn can unite in various ways with still other carbon derivatives and inorganic chemicals to form living tissues.

This capacity to combine, just as with inorganic chemicals, derives from the existence of *valency bonds,* or links, that allow the combinations to take place. These bonds are not physical bindings like cords. They are more like magnetic attractions, although they are not actually magnetic. Each unit of carbon, like any other elemental chemical unit, is an atom, and each carbon atom has four valency bonds that can be represented as follows, although they don't actually occur like this physically:

$$-\overset{\displaystyle |}{\underset{\displaystyle |}{C}}-$$

In a combination, each valency bond must be linked to one valency bond of another atom. All valency bonds must be occupied or the compound will not form. The simplest carbon combination occurs with the gas hydrogen, abbreviated as the letter H. Hydrogen has only one valency bond. The combination then can be represented as:

$$\overset{\displaystyle H}{\underset{\displaystyle H}{H-C-H}}$$

A combination of atoms such as this is called a molecule. A molecule made up of hydrogen and carbon is called a *hydrocarbon.* This particular compound is a molecule of the asphyxiating gas *methane,* sometimes called marsh gas or coal miners' Black Damp. Instead of writing the compound with the valency bonds showing, it is more commonly written CH_4.

Carbon actually has the capacity to link with only a relatively few other chemicals, such as oxygen, abbreviated O; nitrogen, abbreviated N; chlorine, abbreviated Cl; and sulfur, abbreviated S; but it can do so in an extraordinary complexity of chains, loops, and cyclic variants of loops.

For example, if two molecules of methane are united, two of the H atoms drop off and you get *ethane:*

$$\overset{\displaystyle H\ \ H}{\underset{\displaystyle H\ \ H}{H-C-C-H}}$$

14

That structure can be written CH_3-CH_3, or more commonly CH_3CH_3. When a third carbon is added with its hydrogen attachments, you get the fuel gas called propane ($CH_3CH_2CH_3$). As you add more and more carbon atoms with their attendant hydrogen atoms, you get more and more compounds until eventually the structural system becomes unstable and no longer forms.

Instead of combining all the valency bonds with hydrogen, you can combine some of them to get another set of compounds. For example, you can combine chlorine (Cl) with hydrogen to get *chlorinated hydrocarbons.* These compounds include the anesthetic chloroform, ($CHCl_3$), which is also a solvent, and the solvent carbon tetrachloride (CCl_4), which is a liver poison and a dangerous anesthetic in relatively high dosage. These compounds with chlorine are all potentially dangerous.

All sorts of other combinations exist, some of which are determined by the presence of special groupings of atoms in the molecule. One of these groupings, for example, is called the *hydroxyl group.* It comprises an oxygen atom combined with a hydrogen atom. Because oxygen has two valency bonds and hydrogen one, one is left over and is now free to join up with a carbon. The group can be shown as:

-O-H, or -OH

The empty valency bond can link up with an existing molecule to form a new substance. Thus, when a hydroxyl group combines with a chain made up of carbon, hydrogen, and oxygen, you get an *alcohol.* There are many alcohols depending on the type of chain. Ordinary drinking alcohol (ethyl alcohol) can be represented as:

$CH_3 - CH_2OH$, or CH_3CH_2OH

However, if you link the hydroxyl group with the *methane* molecule rather than the ethane, you get the potentially toxic chemical called methyl alcohol, or wood spirit. Similarly, if you add what's called an *aldehyde group* (-CHO) instead of the hydroxyl group, you will get one of a variety of chemicals called *aldehydes,* of which a common one is the gas *formaldehyde* (HCHO), widely used in the manufacture of plastics and glues. This gas can be an irritant and potentially dangerous if inhaled.

Ketones are chemicals containing the *carbonyl grouping* and can be represented as:

$$\overset{|}{-C} = O$$

Note that the connection between the C and the O is a *double bond* to allow for the two valency bonds of oxygen. A common example is the solvent *acetone* (CH_3-CO-CH_2), commonly used in a wide variety of industrial processes and potentially dangerous if inhaled.

The organic acids, on the other hand, such as *acetic acid,* or vinegar, contain a group called the carboxyl group:

$$\overset{\displaystyle O}{\underset{\displaystyle}{\overset{\displaystyle \|}{-C}}}\text{-O-H}$$

It is commonly written as -COOH. Acetic acid can be represented as:

$$\text{H-}\overset{\displaystyle H}{\underset{\displaystyle H}{C}}\text{-}\overset{\displaystyle O}{\overset{\displaystyle \|}{C}}\text{-O-H}$$

It is commonly written CH_3COOH.

Many other significant groupings exist. Three of them deserve special notice because they are a little more common. These include the *amino group,* which is a combination of nitrogen and hydrogen and may be found in plastics and biological chemicals; the *isocyanate group,* which is a combination of nitrogen, carbon, and oxygen, and is used in the manufacture of plastic foam; and the *carbonyl group,* which is a special combination of carbon and oxygen that may occur as an unwanted contaminant in the making of nickel steel and other processes. Each of these can occur in toxic form as will be noted later.

Not all the carbon compounds occur in chains. Some occur in rings. One of the most common of these ring chemicals, and certainly one of the most hazardous, is *benzene,* which can be represented as:

Benzene is the basis of a whole set of industrial chemicals called *aromatic hydrocarbons* that can be derived from the benzene molecule by changing one or more of the hydrogen atoms to form other compounds. They are called aromatic because they have a spicy smell. Exposure can give rise to brain and blood cell damage.

Note, however, benzene with an "e," should not be confused with benzine, with an "i." Benzine, with an "i," refers to petroleum distillates, and is an old term for gasoline. The chemical rings can also link up with chains and other rings to form still more complex compounds. These chemicals are widely used in industry in all sorts of forms from solvents to pharmaceuticals. The list is virtually inexhaustible; hence, thousands of new compounds are added to the workplace every year and many are never investigated to determine the extent of their potential toxicity.

It is not practical here, of course, to attempt to list all the chemicals that can be found in the workplace, far less to describe their effects and toxicity. The list, of course, will vary according to the nature of the work. The chemicals found in paint manufacturing, for instance, will be different from those occurring in a foundry. The nature and occurrence of some of the more common of these industrial chemicals, along with their toxic effects, is examined in Chapter 7.

Meantime, however, just to show that all the bad stuff doesn't necessarily happen on the shop floor, let's look at another cautionary tale, this time in an ordinary office environment.

CHAPTER FOUR

What Polluted the Secretarial Pool?

The law firm of Snodgrass, Snodgrass, Snodgrass, and Peabody was established in town more than 75 years ago, and although the original Mr. Snodgrass is long gone, his grandson Peter carries on the same tradition of sound respectable legal work that was initiated by his grandfather. Not only is he highly respected by his professional peers, he is also highly regarded by his juniors and clerical staff. The law firm has grown over the last 25 years, however, and now comprises some 12 lawyers and supporting staff. Although cherishing the traditions of old buildings, porticos, and marble stairs, Peter, after much painstaking deliberation, recently moved the entire office, lawyers, legal assistants, secretaries and all, into a new prestigious highrise in the town center. The building is palatial, with a spacious entrance hall in mock marble, smooth and silent elevators, and an elegant carpeted foyer leading to the office suites. The decor is stylish, and everything was done to meet the tasteful demands of Peter Snodgrass himself. In particular, the office where the secretaries operate, known irreverently as the pool room, is not only attractive but each workstation was designed to meet the demands of both efficiency and comfort. The room has no windows but is richly carpeted. The walls are panelled with light oak pressboard. The room is 25 feet wide by 40 feet long and provides ample space for the six secretaries in an open plan, with each secretary occupying her own half-walled cubicle. The cubicles themselves are spacious and tastefully finished. Each is open at one end, and of course, each has an open space between the top of the cubicle and the ceiling. Within each cubicle are the usual accoutrements of computer desk, credenza, video display terminal, and other devices such as printer, document holder, and telephone, all designed and laid out in a manner best suited for easy and efficient use. Filing cabinets, personal storage, shelves, and everything else that is needed are all available and of the best quality. At one end of the room are two large copying machines and two fax machines.

The building, of course, is air conditioned, heated in winter and cooled in summer. The ventilation is provided by overhead ceiling vents. The lighting in the pool room is more than adequate. Because the work involves extensive use of video display terminals, eliminating or at least minimizing glare from the video screens is important. That, in fact, was one of the reasons the room did not have windows. Consequently, as advised by a lighting consultant, general illumination was provided by overhead glare-free lighting units at an overall level of 800 lux and enhanced where necessary by individual and adjustable bench lights at each workstation. A *lux,* incidentally, is a measure of the amount of light hitting an area. An ordinary flexible desk light might provide about 400–500 lux.

Everyone was excited and pleased to move into the new office building. The other building was getting cramped, and although it was ornate, and even stately, it was old and not that efficient. The secretaries, in particular, were delighted with their new quarters, their semiprivate cubicles, and their ergonomically designed workstations and seats. Indeed, Peter Snodgrass himself was delighted, not only with his elegant new office, but also with the feeling of high morale that permeated the firm.

And so it came as a shock when, after a month or so of working there, the secretaries one after another began to complain of itching and irritation in the eyes, sometimes in the nose, and even occasional sore throats. As a few weeks went by, the complaints became widespread and the irritation almost continuous.

Mr. Snodgrass, of course, concerned as usual with the welfare of his employees and with lowered morale and reduced productivity, called in his advisors to investigate. At first, it was thought that this might be an ergonomic problem. Ergonomics is a science that deals with the health, safety, comfort, and efficiency of a worker in his/her workplace. A major component of ergonomics, is the study of the relationships that exist between the worker and the "machines" that he or she is working with, or, as it is sometimes called, the person-machine system. In other words: Are those "machines" designed to properly enhance human capacities and minimize human limitations? The "machines" in this case, of course, refer to the layout and components of the workstation, such as the cubicle, the computer desk, the video display terminal, and the printer—all the equipment and tools with which the secretaries come in contact during their day-to-day activities. Could there be something in the organization and layout of the workstation and equipment that might be giving rise to these burning and irritated eyes? Constant staring at an improperly designed, poorly placed, and badly illuminated display terminal might conceivably cause burning and irritation of the eyes, although perhaps not the nose and throat. However, when the ergonomist was called in, she found nothing at all that she could complain about. The workstations were virtually ideal from an ergonomic point of view. Computer desks, credenzas, seats, equipment—all were state-of-the-art design, adjustable where required, properly located and positioned, properly lighted, with no glare and no ergonomic problems of any sort.

What then? Could the problem be an incoming ventilation one like that of Mrs. Madison (although they had never heard of Mrs Madison!)? The hygienists tested out the ventilation system. The incoming air was as pure as a sea breeze. Indeed, it was purer than the air outside the building because it was purified before entering the system. Testing of the air in the room, however, revealed a contaminant was present, a hydrocarbon, in fact. A hydrocarbon, of course, basically is a chemical compound made up of hydrogen and carbon. A problem with testing for chemical compounds, however, is that while often you can show the presence of a *class* of chemicals, it is difficult to identify some particular chemical unless you know what you are looking for. The problem became one of identifying what particular hydrocarbon might be present. None of the secretaries smoked, so the problem couldn't be the effects of tobacco smoke, which is a common cause of problems. Tobacco smoke contains thousands of chemicals, many of which can be toxic. Nor did the problem come from the chemicals used in the copying machine, nor, for that matter, any of the few other chemicals that were used in the office.

Meanwhile, the difficulties in the pool room continued. In fact, they were getting worse, to the extent that one or two of the secretaries had to stay home from work for a day to recover. Interestingly, when they stayed away from work for a day, their problems cleared up, but started again when they returned. So clearly the cause of their problem lay in the pool room itself.

And then one of the operators, a legal secretary named Marlene Leblanc, had a *Thought*. She and her husband had a motor home they used whenever they had the opportunity. The motor home was quite large, 28 feet long, with a cleverly designed living area, kitchen, bathroom, and bedroom that took full advantage of every inch of space. It was panelled inside with wood-grained panelling in the bedroom and other wood or plastic lining elsewhere. And although the motor home was over a year old, it looked shiny and new inside because it had been recalled by the manufacturer for renovation of the interior.

So far as Marlene knew, it had been recalled because of problems with the interior panelling. She was told that many plastics, resins, glues, and even composites such as woodgrain panelling and pressed wood, and, for that matter, some carpets, were either derived from or compounded with the chemical formaldehyde. As already noted in Chapter 3, formaldehyde is a simple example of the class of chemicals known as aldehydes and is made up of hydrogen, carbon, and oxygen organized in a particular fashion. In its original state, formaldehyde is a gas. Sometimes it is dissolved in water to produce formalin and used as a preservative for biological specimens. In industry, however, it is used in the manufacture of plastics, other synthetic resins, leather, rubber, and as a glue for plastics, woods, and even metals. In the manufacture of plastics and other materials, it is first of all compounded with various stabilizing materials and then allowed to "cure," or process, in the presence of heat. Sometimes, however, the curing process is incomplete. Should that happen, and should the other conditions be right, then formaldehyde gas can seep out of the resin, or the glue, or whatever, into the atmosphere.

Formaldehyde is an irritant gas. Even in very small concentrations, it can cause disturbing irritation, particularly in the eyes and the upper breathing passages. As the concentration is increased, the person exposed to the gas begins to have tears in the eyes and, after a few minutes, the discomfort becomes quite profound. If the concentration becomes great enough, the tears become intense, the person starts to cough, and eventually has difficulty breathing. Ultimately, on exposure to high concentration, the victim can get flooding of the lungs with fluid in a condition known as pulmonary edema, which causes very serious injury and even death.

The company that manufactured Marlene's motor home found that, in some cases at least, the panelling and glue that lined much of her cabin was indeed faulty and that formaldehyde gas was being released. There wasn't very much gas being released, but the liability potential was severe. Hence, the recall. It turned out that Marlene's motor home was not affected, but it got renovated nevertheless, much to her delight.

Anyway she had a *Thought*. Could it be that the panelling in the pool room was in some way faulty? Could it even be the carpet? Somewhat hesitatingly she went to see Mr. Snodgrass because, although she was a very good legal secretary, she was certainly no scientist. But he was a good manager and listened carefully. As a matter of fact, any suggestion was welcomed at that time, no matter how far out. So back came the hygienist. Was formaldehyde in the atmosphere? Yes, it certainly was. Not very much, only about three or four parts per million (we'll see what that means in the next chapter), but there was enough to account for all the symptoms. The carpet was fine, but the panelling was at fault. Fortunately for Mr. Snodgrass, this was the only room with panelling. But, of course, all of it had to be replaced, as indeed it was, with much disruption and great relief.

And what happpened to Marlene? She was promoted and became Mr. Snodgrass's private legal assistant.

CHAPTER FIVE

What Happens to Toxic Chemicals in the Body?

HOW DO CHEMICALS GET INTO THE BODY?

Chemicals can get into the body in three basic ways: by breathing them in (inhalation), by swallowing them (ingestion), and by absorbing them through the skin (transcutaneous absorption). Let's look at each of these in turn.

INHALATION

It's easy to understand that if you breath air containing gases, mists, fumes, or whatever, you will inhale these materials along with the air. But what happens then? Normally, you inhale by way of your nose, although if your nose is blocked, you can inhale through your mouth. Inhaling through your mouth is not so good because your mouth is not designed to filter out impurities. Your nose on the other hand, is lined with hairs that act as fine dust filters and it is also connected with the sinuses that can act as dust collectors. The sinuses are bony caverns in the front and forehead portion of the skull. They are very useful, even though they sometimes get inflamed in the condition called sinusitis. In particular, they serve two functions. They warm, or sometimes cool, the air to bring it to body temperature, and they moisten it to bring it to the humidity level of the air passages. The temperature and humidity render the air compatible with conditions in the lungs. The sinuses also collect some of the larger dust particles. Not so long ago, and sometimes even on rare occasions today, persons working in cotton or textile mills had to breathe air so thick with cotton dust they couldn't see across the room. In that situation, their sinuses could get totally blocked with cotton dust in the disabling condition called byssinosis, which we'll encounter again in Chapter 9.

From the sinuses, the inhaled air then passes on down the air passages into the lungs. In the lungs, an exchange takes place between the gases carried in the blood and the gases in the air. The blood gives up carbon dioxide, which is manufactured in the living processes of the body tissues, and takes on oxygen from the air to allow these processes to take place. Meantime, of course, the blood also absorbs the various other gases and vapors that might be present. The dust that has not been filtered in the nose or blown out when you exhale begins to collect in the finest air passages, where, as we shall see in Chapter 9, it can give rise to serious problems.

INGESTION

While it is easy to understand how readily you can inhale chemicals in the air, it is perhaps not so easy to understand how you can swallow them. You certainly don't go around eating toxic chemicals. What happens, however, is that chemicals get on your hands, on your gloves, or on your face, and you then transfer them accidentally or unthinkingly to your mouth. If you're working all day with lead, perhaps without gloves, and you rub your mouth, or light a cigarette, or eat a sandwich without first carefully washing your hands, you can easily transfer some of those fine lead particles to your lips and then to your mouth. Doing it once doesn't matter. But doing it a dozen times a day, every day, does.

Once the substance gets into the mouth, it's on its way into the body. It passes down to the stomach and into the small intestine. There it is picked up by the blood stream through the wall of the intestine and passed into the liver where it can be transported throughout the whole body.

TRANSCUTANEOUS ABSORPTION

Transcutaneous absorption, or absorption through the skin, can occur only with certain types of chemicals. The skin is really a remarkable organ. Note that it is an organ, not just a piece of soft protective leather. It has many functions besides holding all the muscles, bones, and organs in place. The skin, of course, is a sense organ, but is also one of the main elements in heat control, partly because it permits sweating and partly because it permits the direct exchange of heat with the environment. Along with the kidneys, it is involved in maintaining the fluid balance in the body through sweating and, to some extent, in excretion or removal of unwanted substances in sweat. Some metals, for example, like arsenic, can be excreted in the sweat or even in the hair.

As a protector, however, the skin is especially useful. In particular, it can prevent most chemicals from passing through it, partly because of its mere physical presence, and partly because of its complex layered structure. Within the skin is a fatty layer that prevents the passage of chemicals like common salt, which are soluble in water but not in fat. Most soluble chemicals are soluble in water but not in fat or oil, and consequently, they can't get through the skin. Unfortunately, there is a class of chemicals that are used for the very purpose of dissolving oil, grease, and fat. These chemicals are the solvents, such as commonly used acetone, hexane, or Stoddart solvent. Because these solvents can dissolve fat, they can, in fact, pass right through the skin and, in doing so, not only damage the skin but also get into the body. Solvents are discussed in Chapter 7. Solvents, of course, are not the only substances that can pass through the skin. There are numerous others; two of the most notorious are lead and mercury. These also are discussed in Chapter 7.

WHAT HAPPENS TO THE TOXIC CHEMICALS ONCE THEY'RE IN?

Perhaps a little surprisingly, there are really only two basic actions that foreign substances can have on body tissues. These are interference with the function of these tissues and damage to and/or destruction of them.

INTERFERENCE WITH FUNCTION

All body tissues are made up of cells. These cells are differentiated for special purposes. Thus, there are brain cells, and muscle cells, liver cells, and kidney cells, and so on. The cells bond to form tissues, and the tissues combine to form organs such as the kidney, or body parts such as the muscles. All these tissues are active, and each has a function. The presence of a toxic substance can modify that function. It can't change the nature of the cell. You can't change a liver cell into a brain cell. But a toxic substance can destroy a cell or increase or decrease its normal activity. Stimulation of cell function, that is, increase in activity, will cause over function; depression of cell function will cause reduced function. Either of these activities, if carried to extremes, can lead to dysfunction or failure, with various manifest illnesses or even death. Thus, for example, the gas acetylene, or for that matter various other gases, acts as a depressant of cell function and can reduce brain activity to the point of loss of sensation and unconsciousness. If the exposure is sufficiently mild, as for example in drinking a shot or two of bourbon, there will be a later recovery with little or no residual effect. The greater the exposure, the greater the recovery time and the more severe the recovery symptoms, or as we might call it "the hangover." If the exposure is very severe, however, or moderately severe but frequently repeated, there could be death or permanent damage. This characteristic of loss of sensation and unconsciousness, by the way, is exploited in the use of surgical anesthetics such as chloroform or ether.

Alcohol is a special case. Alcohol is not a stimulant as is popularly supposed. It is a depressant. It depresses the function of the brain cells in a sequential manner, acting first on those "higher level" functions of judgment and self-criticism so that one no longer sees one's conduct as being foolish or inappropriate. Thus, you can do crazy and even stupid things and think you are enjoying yourself. You then take more alcohol to get an even greater effect, but now the alcohol begins to take effect on other parts of the brain until eventually you become unconscious.

There are also industrial chemicals that can *increase* cell function. Mercury, for example, can cause severe destruction of tissues, but it also acts as a stimulus to cell activities and, in particular, to brain cells, causing various manifestations of trembling and uncoordinated movements as well as mental excitement and agitation.

About 100 years ago, when people used to wear felt hats, the hat maker used mercury to process the felt. He applied the mercury to the inside of the hat brim by dipping his fingers into a pool of mercury and then rubbing it along the felt. Over the years, and often not before too long, the hat maker absorbed enough mercury in this manner to get mercury poisoning. He would then show all the symptoms of agitation and so on that gave rise to the phrase "mad as a hatter," and to the character of the Mad Hatter in *Alice in Wonderland.*

TISSUE DAMAGE AND DESTRUCTION

Certain substances, and particularly chemicals that are strong acids or corrosives, act to damage and destroy tissues rather than simply interfere with their function.

Not all destructive chemicals are acid or corrosive, however. For instance, although exposing the skin to strong acids, such as muriatic, or alkalis, such as lye, will result in immediate damage and ulceration, application of less irritant chemicals, such as benzene, gasoline, and various solvents to the skin can produce a response that is really an attempt by the skin to protect itself. This response initially is dryness and scaliness and then proceeds on until the skin becomes hard and horny with the development of fissures and cracks. At any time during this process, the conditions can be aggravated further by the occurrence of infection, with inflammmation (redness, swelling, and pain) and ulceration.

If the irritant material is swallowed, particularly if it is corrosive, the destruction, of course, takes place in the digestive tract, again with inflammation and ulceration, and sometimes indeed with gross destruction. Similarly, inhalation of irritant chemicals can cause destruction of tissues in the breathing passages. Normally, exposure to an irritant gas will produce an immediate bout of coughing and spluttering that will cause you to run from the area if you can. If you can't get away, however, and if the irritation is severe enough, for instance an exposure to a high concentration of chlorine or nitrogen dioxide, the destruction will lead to a protective outpouring of fluid into the lung called pulmonary edema that can fill up the lung to the extent that the victim will drown in his own fluids. If the irritation is less severe, it can still cause destruction of lung tissue, but, in this case it will later result in the development of bronchitis and disabling lung diseases.

There is still another way in which chemicals can cause damage to internal tissues. When chemicals are absorbed into the blood stream, they are passed to the liver, as all body chemicals are, where they can undergo chemical processing. However, sometimes in processing toxic chemicals, the liver itself may be damaged to a greater or less extent. Most people are aware, for example, that excessive and prolonged consumption of alcohol will cause liver damage that can lead to death. As another example, carbon tetrachloride has major effects on the brain cells, but in the process of being processed by the body, it can cause significant damage to the liver, with resulting liver disease. In yet another fashion, mercury is excreted by the kidney after it exerts its widespread effects on the body. It is not excreted without cost, however, for in so doing, the very tissues of the kidney become progressively destroyed.

COURSE OF EVENTS WITHIN THE BODY

As we've seen above, there are really only two basic ways in which toxic chemicals can affect body tissues and cells. There are, however, different ways this happens. A chemical, once in the body, can exert an immediate effect and then be eliminated, or it can be allowed to accumulate to a particular threshold level before it takes further action, or it can be changed in form to a less toxic state.

IMMEDIATE ACTION AND ELIMINATION

This is the kind of action we already noted when we described what happens when a person inhales a toxic gas such as chlorine or oxides of nitrogen. The effect is immediate, and the elimination takes place as the effect is being produced. Most chemicals, in fact, begin to be eliminated while they are still being absorbed or even while they are being changed to some other form. Carbon monoxide, for example, will get into the bloodstream as soon as it is inhaled and attach itself to the hemoglobin in the blood cells that carries the oxygen. In doing so, it will block the ability of these blood cells to carry oxygen. Depending on the concentration of the carbon dioxide, the victim can then be dopey, unconscious, or dead within minutes to hours. Common industrial gases such as chlorine or ammonia will cause immediate irritation as soon as they are inhaled, even while they are being breathed out again. Solvents like chloroform and acetone are carried into the bloodstream and up to the brain where they will give you an almost immediate "high" if the concentration is great enough. Unfortunately, the margin between "high" and unconscious is very narrow, and the "high" can lead to dizziness, unconsciousness, and death. However, even as these substances act, they are destroyed in the liver and eliminated in the kidney. This type of action, however, as we'll see in the next section, does not apply to all chemicals.

ACCUMULATION TO THRESHOLD AND SUBSEQUENT ACTION

Some chemicals pass into the body with little or no immediate effect, and only produce significant effects years later. One of the most important of these is lead. When you take in small quantities of lead, day by day, by ingestion, or through the skin, you may have little or no problem other than perhaps some mild colic or stomach ache, perhaps not even that if the daily dosage is small. The lead that you take in, however, passes into the bloodstream and is carried throughout the body. For some reason, not clearly understood, it then settles largely in the bones, particularly the long bones such as the arm and the leg, and there it stays year after year as more and more lead is accummulated.

Eventually, however, there comes a time, perhaps ten or twenty years down the road, when the bones can hold no more. The incoming lead is still passed into the bones, but, to make room for it, lead must now pass out from the bones to re-enter the blood stream where it rapidly begins to accumulate to a level at which it exerts toxic effects on other body tissues. Lead, in fact, is a widespread body poison, affecting in one way or another practically every type of body tissue. Most of the effects, however, are felt in the digestive tract, in the red blood cells, and in the nerves that supply the muscles. The presence of the lead gives rise to such symptoms as very severe abdominal colic or griping from lead in the digestive tract, weakness and lassitude associated with the blood cell damage because the blood can no longer carry enough oxygen, and wrist or foot paralysis (wrist drop, foot drop) associated with the damage to the nerves.

DETOXIFICATION AND/OR ELIMINATION

Ultimately, over periods that may be very short, as in the case of volatile substances such as acetone or alcohol, or very long, as with substances such as the dyestuff aniline which latches on to the hemoglobin in the blood that normally carries oxygen, most chemicals will be removed from the body by the body's natural processes. As we have seen, however, there are exceptions, such as lead, that will stay in the bones forever.

Elimination takes place largely by way of the kidney, which acts like an active filter for all unwanted body substances, although sometimes it gets damaged in the process as we have already noted. Some elimination of ingested material that has not been absorbed also takes place by way of the bowel, or even, in the case of volatile materials such as alcohol, by way of exhalation of air. Everyone knows the smell of excreted alcohol on the breath of someone who has been drinking. Some materials, such as mercury, can be excreted in part through the skin or, for that matter, the saliva, and indeed a substance such as arsenic can even be excreted by being incorporated into the hair.

Not all substances, however, are eliminated in the form in which they are acquired. As was previously noted, all the blood in the body passes through the liver. The liver acts as a chemical processing factory. It takes the wanted chemicals from foodstuffs and transforms them into chemical building blocks for use by body cells in building the materials they need for their structure and function. It also takes the body's chemical waste products and prepares them for elimination. Thus the liver is capable of undertaking many different chemical processes.

Consequently, should a foreign chemical arrive in the liver from a toxic industrial source, for example, it can be exposed to any or all of these processes. Sometimes a chemical reaction will result; more often it won't. If it does result, then the foreign chemical is changed and is eliminated in a changed form. If that change makes the chemical inherently less toxic, then it is said to have undergone *detoxification.* For example, acetic acid goes through a relatively simple form of change in the liver. Chemical catalysts called enzymes act on it to add oxygen to the molecule and then break down the newly formed molecule into the gas carbon dioxide, along with some water. The carbon dioxide is then removed by exhalation through the air passages, and the water is removed by the kidney. The solvent, ethyl acetate, is also broken down in the liver. It is first broken down into two substances, namely, acetic acid and alcohol, which are each subsequently further broken down into carbon dioxide and water, again by the action of enzymes. There are many other varieties of reaction that can take place. Unfortunately, only a few chemicals can be said to be capable of detoxification, and often only some of a given chemical is detoxified and the remainder goes on to cause problems.

Sometimes, however, although rarely, a chemical may change before it gets into the body, perhaps with disastrous results. So let's look at a somewhat unusual case that occurred under conditions that were it not for unexpected circumstances, would normally have been considered acceptable.

Harry Hemmler's Unlucky Day

The day the plant manager decided to put a curtain around Harry Hemmler's workstation was Harry Hemmler's unlucky day. The company he worked for over the last five years made metal wheels for wheelbarrows. It was a small company, with only six shop floor employees. Labor inspectors and any others who might be interested in health and safety largely left them alone. The employees were not unionized, but were paid reasonably well and were glad indeed to have a tolerable job during some fairly hard times. Because of its size and isolation, however, and because of relative indifference on the part of management, health and safety practices were minimal and more or less at the whim of the owner. As a result, even though there wasn't much chance of exposure to toxic materials, it was a noisy place with metal presses, grinding wheels, and other equipment, and generally poor health and safety conditions.

The wheels were stamped and cold forged out of sheet metal. In the course of their transformation from sheet metal to the shape of a wheel, they were liberally coated in cutting oils and other lubricants. It was necessary, of course, to remove these greasy substances before the wheels were subjected to heat treatment. Harry's job was to take each rim as it came to him suspended on a hook on an overhead track, remove it from the hook, and dunk it in a vat of carbon tetrachloride degreasing liquid. When the process was complete, he would then remove the rim from the vat and hang it back on the rack for onward transmission.

The vat wasn't very large, about 48 inches wide, 42 inches high, and 36 inches front to back. It was kept filled with carbon tetrachloride up to about 6 inches from the top. The top was supposed to be covered by a hinged lid when the vat was not in use, but more often than not, the lid was left off. The company owner learned some years previously that carbon tetrachloride was a dangerous chemical, and because it vaporizes very readily he had installed a 26-inch fan above and at the back of the vat to direct the vapors through the wall to the exterior. In fact, the fan wasn't all that effective. Although Harry didn't know it, nor would even have known what the numbers meant, the concentration of carbon tetrachloride at his nose and mouth level was later found to vary from about 10 to 20 ppm, depending on circumstances. As we shall see, the permissible acceptable threshold limit for carbon tetrachloride is 5 ppm, so while he worked there, which could be from 4 to 8 hours per day every work day, Harry was certainly exceeding the recommended level. But that wasn't his problem, and, in fact, he did not particularly complain about the exposure because it sometimes gave him a bit of a "high" like he used to get breathing glue in a paper bag as a kid. He didn't know, however, that the exposure was giving him some other problems that could affect him later on.

Carbon tetrachloride was at one time widely used in industry, and still is to a certain extent, although its use as a degreaser or common solvent has been banned in many jurisdictions. Until recently, however, it was commonly used as a degreaser and a solvent for oils, fats, lacquers, varnishes, rubber waxes, and resins. Two of its most common uses were in the dry cleaning industry and in portable fire extinguishers.

Its effects on the body are widespread. Inhalation of the vapor is the most common mode of entry, and when the chemical is inhaled in sufficient concentration, it has an immediate effect on the brain (hence the "high" that Harry was partial to) and ultimately, over longer periods, it causes damage to the liver and kidneys. Now it so happened, that although Harry had been working for five years with concentrations of carbon tetrachloride above the recommended limit, any obvious damage to his brain, liver, and kidneys at that time was no more than he would have derived from knocking back a six pack of beer several times a week. So what was the trouble?

Well, as I said at the beginning, Harry's unlucky day began when the plant manager decided to put a curtain around Harry's workstation. In fact, in a sudden burst of health consciousness, the manager decided to isolate Harry's station and give Harry some respiratory protection while he was working with the carbon tetrachloride. And for this we must congratulate him, although it was not the best way to control the situation. Ideally, he should have replaced the carbon tetrachloride with some less toxic degreaser, of which many are available, and then automated the degreasing process in such a manner that it didn't require personal handling. Anyway, he decided to put a curtain around the workstation. To do this, he had to weld in position some supports and rails on which to hang the curtain. And this he did during the normal working day while Harry was still dunking away with his rims.

So far, so good. Harry worked all day, with his usual coffee and lunch breaks, and finally got home about 4 o'clock in the afternoon. It was his bowling night, and he went bowling about 7 p.m., getting home about 10:30, ready for bed. And as he climbed into bed, he said to his wife, "You know, I think I'm getting a cold. I've got a bit of a sore throat and a cough coming on."

By midnight, his chest was hurting, his cough was severe, he had some difficulty in breathing, and he was bringing up lots of frothy phlegm. He was in the hospital by 1 a.m., in intensive care. It was two months before he returned to work.

So what was it all about? Well, poor Harry was the unfortunate victim of a curious chemical reaction. In the manner shown in Chapter 3, the chemical formula for carbon tetrachloride can be represented as this:

$$
\begin{array}{c}
\text{Cl} \\
| \\
\text{Cl–C–Cl} \\
| \\
\text{Cl}
\end{array}
\qquad (CCl_4)
$$

When you look at that formula, you will see that, as the name implies ("tetra" is Greek for 4), carbon tetrachloride is made up of one atom of carbon (C) and four atoms of chlorine (Cl). Now, it so happens that in the presence of ultraviolet light, two of the chlorine atoms can become replaced by oxygen from the atmosphere to make a compound that can be represented like this:

$$
\begin{array}{c}
\text{Cl} \\
| \\
\text{C=O} \\
| \\
\text{Cl}
\end{array}
\qquad (COCl_2)
$$

And that compound goes by the name of phosgene. You may have heard that phosgene is a war gas that was used with devastating effect in World War I. It was never used in World War II, although supplies were available. It is said, however, that it has been used since. Phosgene is a toxic gas that acts as a very severe irritant to the lungs and respiratory passages. It has a curious characteristic, however. It does not take effect right away. It gets breathed into the whole respiratory system, of course, right down into the lungs, and there it gets absorbed into the tissues, where, over a period of time, the chlorine atoms are broken loose from the phosgene molecule. It's the resulting chlorine that causes the severe irritation, from inside, as it were. And that's why Harry didn't get into trouble until many hours after he left the site of the exposure.

But where did the ultraviolet light come from? There lies the unlucky part. Welding, of course, generates ultraviolet light, and if the ultraviolet light is in line of sight of the carbon tetrachloride, then the reaction can take place. And, of course, carbon tetrachloride vapor was present most of the time in fairly high concentration above the vat where Harry was working and where he could breathe it in his usual fashion. But poor Harry, who never had too much obvious trouble from his exposure to carbon tetrachloride, found himself on this occasion quite unwittingly breathing phosgene. He probably didn't notice the difference. To the trained expert, phosgene is said to have a smell like new mown hay, but, of course, Harry was also breathing the sweet smelling carbon tetrachloride, which to him was really not too unpleasant.

This is not a common situation on the ordinary shop floor, if only because you don't see too much carbon tet around anymore. But it certainly does illustrate the saying that things ain't always what they seem—even in a small manufacturing plant.

CHAPTER SEVEN

What's in the Air?

Perhaps the title of this chapter should be "What's in the Air, and on the Ground, and on Your Skin, and in Your Mouth, and in Your Lungs, and Wherever Else You May Care to Name." Chemicals in the workplace can be anywhere and everywhere and take many forms. The general term applied to particulates of different types in the atmosphere is an *aerosol*. Aerosols are made up from dusts, fumes, smoke, gases, vapors, and mists. These terms are often misused and, consequently, it is wise to have an understanding of what each one means.

Dust is made up of solid particles derived from such activities as crushing, grinding, or other abrasion of solid materials. Fumes are solid particles condensed from the gaseous state of a metal, such as, lead oxide from smelting, iron oxide from welding, and zinc and magnesium from grinding. Smoke is made up of particles of carbon or other sooty products derived from incomplete combustion of some material. A gas, according to its formal definition, is a fluid that at the specific temperature of 25° Celsius and an atmospheric pressure of 760 mm of mercury completely occupies an enclosed space. A vapor is the gaseous state of a substance normally found in the solid or liquid state that has been changed because of change in temperature or pressure. A mist is a finely divided liquid suspended and dispersed in the air. It is generated by condensation from gas to a liquid state or by break up of a dispersed liquid. In industry, it is commonly found to exist as an oil mist used in cutting and grinding oils. In practice, it doesn't make a great deal of difference what name you apply to these aerosols, but it can cause problems when you want to be specific.

The gases and the fumes and the vapors and the mists, of course, tend to float in the air along with fine dust particles. And from there, they get into your nose and your breathing passages, into your mouth and digestive system, and, of course, into your eyes and on to your skin. The liquids and solids, on the other hand, are normally contained, unless they are spilled, and they more commonly come in contact with your skin, unless you transfer them to your mouth and other areas by way of your hands. We have already seen in general what happens after they get on the skin and into the body. In this chapter, however, we are going to look at some of the more common specific chemicals and see what their effects might be. One way to do this would be to list all the appropriate chemicals alphabetically and then describe their effects. This would be quite effective, but it is probably more useful to classify them in groups so that the understanding becomes more meaningful. Unfortunately, as was noted in Chapter 1, toxic chemicals do not divide themselves conveniently into groups according to their structure or function, with each class having the same type of effect. As we shall see, this does occur to some extent, but not cleanly enough to make the classification useful. Consequently, the grouping that is proposed here is a mixture of physical, chemical, and functional characteristics that, although perhaps not academically pure, will make the subject easier to comprehend. So let us look at this grouping and see if we can get a reasonable understanding of the kinds of chemical we are likely to encounter in the workplace. But remember, this classification is not complete, and the chemicals presented here represent only a small number of the multitude of chemicals that can be found. The actual chemicals you can encounter will depend very much on the nature of the work being conducted in your particular workplace. With that in mind, let's look at the chemicals in the following categories:

- Irritant gases
- Choking (asphyxiant) gases
- Solvents
- Heavy metals
- Metal fume fever producers
- Chemicals that damage the blood cells
- Chemicals that cause sensitization

IRRITANT GASES

Irritant gases, of course, produce their primary effect when you breathe them into your air passages and lungs. The effect is one of immediate and commonly, violent irritation, with coughing, spluttering, gasping, a burning sensation and pain, while, to the extent that you can, you stop breathing in self-protection.

Some of the more common of these gases are listed in Table 7-1. The Hazard Potential column gives the permissible exposure limit or PEL (see Chapter 16) in terms of parts per million (ppm), which refers to the concentration of the material in so many parts per million parts of air, or milligrams per cubic meter of air, which is another method of saying the same thing.

Table 7-1
List of Selected Irritant Gases

Gas	Usage	Hazard Potential and PEL
Ammonia	Refrigeration, manufacturing	Moderate (25 ppm)
Bromine (vapor)	Bleaching, gold extraction, anti-knock gasoline	Very high (0.1 ppm)
Chlorine	Metal fluxing, sterilization, bleaching, manufacturing	Very high (0.5 ppm, 1989*)
Formaldehyde	Manufacturing, fiber board, plastics	Very high (0.75 ppm)
Hydrogen chloride	Chemical intermediate, steel pickling	High (TLV ceiling 5 ppm)
Hydrogen fluoride	Glass etching, removal of sand from castings	High (3 ppm)
Hydrogen sulfide	Byproduct in petroleum processing	Moderate (10 ppm)
Oxides of nitrogen	Welding, manufacturing, silo gas	High (TLV 3 ppm)
Ozone	Arc welding, water purification	Very high (0.1 ppm)
Phosgene	Manufacturing, metallurgy	Very high (0.1 ppm)

(continued on next page)

Table 7-1 (continued)
List of Selected Irritant Gases

Gas	Usage	Hazard Potential and PEL
Phosphine	Fumigation, manufacturing	Very high (0.3 ppm)
Sulfur dioxide	Manufacturing	High (2 ppm, 1989)
Sulfuric acid	Bleaching, food processing	High (1 mg per cu meter)

The date 1989 refers to the revised (and not approved) version of the original PEL of 1971.
Source: multiple texts.

As you might imagine, the more irritant the gas or the stronger the concentration, the greater are the symptoms. Normally you can protect yourself with personal protective equipment or you can immediately remove yourself from the source by running to fresh air. However, if you're caught unprotected in a situation from which you can't get away, and you continue to breathe, then the gas gets down into the lungs and causes violent irritation in the soft tissues of the lungs. The result is an outpouring of protective fluid into the lungs in what is referred to as *pulmonary edema.* This outpouring in itself may be sufficient to choke you. Normally, however, if you get a relatively mild exposure in the upper air passages, you will be left with a few days of local irritation of these passages in the form of, for example, *pharyngitis,* which is inflammation of the back of the throat, or *bronchitis,* which is inflammation of the air passages themselves.

You can, however, get a much more severe aftereffect, and if the irritation progresses as far as the lungs and you survive the pulmonary edema, you will get an intense inflammation of the lungs in the form of chemical pneumonitis, a form of pneumonia, which in itself can be fatal. Similarly, if you spend a prolonged period breathing very low concentrations of such gases, as can happen in industry, you will, of course, get continued irritation of these passages with possible damaging effects.

CHOKING (ASPHYXIANT) GASES

Asphyxia means deprivation of oxygen. As you well know, if you are deprived of oxygen for more than a few minutes, you die, and whether that deprivation comes because someone is choking you with a pillow over your head or because some other gas is blocking out your oxygen, you will still be deprived. Normally one fifth, or 20 percent, of the air we breathe is made up of oxygen. The remainder is almost entirely nitrogen. As we increase the proportion of gases other than oxygen in the atmosphere, we automatically reduce the proportion of available oxygen until it drops below a value sufficient to sustain life. We can get away fairly comfortably with about 15 percent of oxygen, but, as the amount gets less, we begin to get into serious trouble.

The choking gases, then, are not irritant. In fact, they are normally harmless, except that their mere presence can block out life-giving oxygen. Three of the common choking gases are nitrogen, carbon dioxide, and methane. Others include acetylene, hydrogen, liquified petroleum gas (LPG), and refrigerant gases. They all have the same ultimate result, namely, loss of consciousness and death if the exposure is sufficiently prolonged or the proportion in the atmosphere is sufficiently high.

Of the various gases noted above, nitrogen and carbon dioxide are part of the air we breathe, as are argon, neon, and helium in minute quantities. Nitrogen, in fact, as noted in the previous paragraph, forms some 80 percent of our breathing air, and it is only when by some chance the percentage increases above 85 percent that it begins to act as a choking gas, the more so as the proportion gets higher. Curiously, however, nitrogen can have another serious effect when we breathe it under pressure, as, for example, when we breath ordinary air in deep water scuba diving. In those circumstances, the air, with its various constituents including nitrogen, is under considerable pressure. When nitrogen is breathed under pressure, it acts as a narcotic, or, in other words, it acts to depress the function of the brain cells in the same manner as alcohol. It has been said facetiously that each 100 feet of water has the same effect as one martini. The effects of nitrogen narcosis, as it is called, begin to become serious at about 300 feet under water. However, that phenomenon has nothing to do with the normal workplace and is mentioned here only out of interest.

We also breathe carbon dioxide every day of our lives. Carbon dioxide is manufactured in the body during the process in which the body cells derive energy from the sugar-like substances known as carbohydrates. It is carried to the lungs in the blood and exhaled every time we breathe. Normally, it is totally harmless. In fact, it is only when the concentration of the carbon dioxide in the breathing air is such as to reduce the proportion of available oxygen that a problem arises. This can happen if, for example, you are in a sealed space where the carbon dioxide from your own body builds up in the atmosphere, or if, for some unknown reason, you are breathing with a bag over your head. The problem of simple choking, however, is complicated by the fact that the amount of carbon dioxide in the body is used by the body to control the rate and volume of breathing. As the concentration of carbon dioxide rises in the blood, the rate and volume of breathing increase. While this feeling may be uncomfortable, it is not harmful until it gets to extremes. However, because of this, the acceptable concentration of carbon dioxide in the atmosphere is generally considered to be about 0.5 percent, or 5,000 ppm.

Methane gas is in itself inert. It occurs naturally as marsh damp in swamps or, more ominously, Black Damp in mines. In addition to being an asphyxiant when the quantity blocks the available oxygen, it is also highly explosive and, hence, particularly dangerous in mining situations.

SOLVENTS

Solvents are commonly used for degreasing metal objects prior to processing as well as for cleaning purposes and in solution with other chemicals. There are many kinds in various chemical forms. The effects of some of these are examined below.

While solvents vary considerably in their effects, certain broad conclusions can nevertheless be drawn. In low concentrations, over months or years, the effects of skin contact with solvents are commonly the result of damage to the fatty layer of skin. As already noted, the skin is a complex organ composed of several layers of different types of cell. One of these layers is made up of special cells containing fat. Because of this, watery substances, and particularly those containing dissolved chemicals, cannot readily pass through the skin. With long exposure to solvents, however, the outer layers of the skin, including the fatty layer, are damaged. This damage gives rise to a form of dermatitis, or inflammation of the skin, which is characterized by dryness, scaly hardness, and even fissures. The skin can become red, inflamed, and painful. Because the fatty layer is also damaged, and may even be destroyed, other toxic materials that previously could not pass through the skin now get a much readier access to the body.

Vapors from the solvents or actual contact can also cause irritation of the eyes and the lining, or mucous membrane, of the nose and throat. If the vapors penetrate further into the respiratory passages, these too will become irritated and cause an irritation of the back of the throat (pharyngitis), or worse still, inflammation of the bronchi (breathing tubes) in the condition of bronchitis, which we have already seen occurring with the irritant gases.

In addition, as we shall see again later, some solvents such as benzene, when contacted over prolonged periods, have the special capacity of acting to destroy the bone marrow that manufactures red blood cells or to stimulate the production of white blood cells in the form of a type of blood cancer known as leukemia. And still others, such as hexane and methyl butyl ketone (MBK), have the capacity to damage the nerve supply to the muscles of the legs, in particular.

On the other hand, if the exposure to a solvent occurs due to inhalation of the vapors in a relatively high concentration for a short period, the effects tend to be found more commonly in what is known as depression of the central nervous system (CNS). Depression of the CNS refers to a reduction in the ability of the brain cells to function. This can take various forms. One example is the excitement and feel-good effects of alcohol or glue sniffing, which result from a mild depression of the CNS that reduces one's capacity to exercise rational judgment. The same alcohol or glue-sniffing in a higher concentration, however, or any other CNS depressant for that matter, will depress the CNS function still further and give rise to such effects as dizziness, drowsiness, loss of coordination, confusion, nausea, vomiting, and pain in the abdomen, accompanied or followed perhaps by various disturbances of heart rhythm, convulsions, unconsciousness, and ultimately death. Fortunately these worst effects occur only at very high concentrations, beyond those normally found in industry, and are rare. But they can happen.

When solvents get into the body, most commonly by inhalation of vapors, they are passed by the blood stream to the liver, which tries to destroy them. Unfortunately, in doing so, the liver itself may be damaged. Most people are probably familiar with the ultimately fatal liver, damage known as cirrhosis that can occur from prolonged severe ingestion of alcohol. The same kind of damage can occur from inhalation of solvent vapors, particularly those of the group that include chloroform and carbon tetrachloride. When the liver has done what it can, the solvent or its derivatives are then passed on by the blood stream to the kidney for excretion. The kidney, too, can be damaged in the course of this process.

Table 7-2 lists some of the more common solvents, along with some of their effects and their hazard potential. The word *acute* in the table refers to a short-term, relatively high concentration exposure, while the term *chronic* refers to a long-term, relatively low exposure. The column labeled Hazard Potential gives the recommended exposure limit. A question mark after a comment indicates that there is some doubt about the conclusion.

Table 7-2
Selected Solvents and Their Effects

Name	Primary effect	Other effects	Hazard Potential and PEL
Alcohol (ethanol)	Irritation of eyes, mucous membranes	CNS depression at higher levels	Mild (1000 ppm)
Amyl acetate	Irritation of eyes, nose, throat, skin	(High exposure) CNS depression	Mild (TLV 125 ppm)
Acetone	Irriation of eyes, mucous membranes	CNS depression at higher levels	Mild (750 ppm)
Benzene	*Acute:* CNS depression	*Chronic:* bone marrow damage. Skin irritation. Cancer?	High (1 ppm)
Butanol (n-butanol)	Irritation of eyes, mucous membranes	CNS depression at higher levels	Moderate (50 ppm)
Butyl cellosolve (Dowanol EB)	Irritation of eyes, mucous membranes	Blood cell damage?	Moderate (25 ppm)
Carbon tetra-chloride	CNS depression	Liver, kidney damage, skin irritation. Cancer?	Severe (2 ppm, 1989*)
Cellosolve (Dowanol EE)	Skin irritation	Eye irritation	Moderate (5 ppm)
Ethyl acetate	Upper respiratory irritation	Sensitization of skin, mucous membranes (rare)	Mild (400 ppm)
Ethyl ether	Eye and upper respiratory irritation	CNS depression at high levels	Mild (400 ppm)
Hexane (Skellysolve B)	Upper respiratory and skin irritation	CNS depression *Chronic:* nerve damage	Moderate (50 ppm, 1989)
Methyl butyl ketone (MBK)	Nerve damage	Irritation of mucous membranes at high levels, also CNS depression	Severe (5 ppm, 1989)

(continued on next page)

Table 7-2 (continued)
Selected Solvents and Their Effects

Name	Primary effect	Other effects	Hazard Potential and PEL
Methyl ethyl ketone (MEK)	Irritation of eyes, mucous membranes, skin	CNS depression at high levels	Mild (200 ppm)
Tetrachloro-ethylene	CNS depression, numbness of extremities	liver damage, skin irritation	Severe (5 ppm)
Stoddard solvent	Mild CNS depression	Irritation of mucous	Mild (100 ppm, 1989)
Toluene	CNS depression	Irritation of skin	Moderate (100 ppm, TLV 50 ppm)
Trichloro-ethylene (Trilene)	CNS depression, respiratory irritation,	Injury to heart, digestive system, liver, kidneys	Moderate (50 ppm)
Xylene (Xylol)	Irritation of eyes, mucous membranes, and skin	CNS depression at high levels	Moderate (100 ppm)

*The date 1989 refers to the revised (and not approved) version of the original PEL of 1971.
Source: multiple texts.

HEAVY METALS

The term heavy metals is rather broad, relative, and nonspecific. There are many "heavy" metals. In industrial toxicology, however, the term is used to describe certain heavy metallic substances that can be distinguished from other metals because of their particular toxic effects. Of these, the most common are arsenic, lead, and mercury. These are the ones that will be dealt with here.

ARSENIC

Arsenic was the popular poison of the Middle Ages and perhaps even of the infamous Rasputin, the Mad Monk, who is said to have used it against the family of the Czars in pre-revolutionary Russia. It was used not only because it was effective, but also the poisoner could drink the same poisoned wine with little effect because he or she could become relatively immune by continued ingestion of small but increasing quantities over a long period of time. However, ingestion of repeated small quantities of arsenic is not a practical method of control in industry.

Arsenic is a steel gray metal. Arsenic and its compounds have a very high hazard potential. They are used in metallurgy, in paint production, in the manufacture of certain types of glass, and in insecticides, fungicides, and rat poisons. In contact, arsenic compounds are irritating to the skin, mucous membranes, and the eyes. The greatest danger, however, is by ingestion. When used as a poison, of course, the concentration is extremely high. In industrial usage, where the concentration is much lower, continued ingestion can, nevertheless, give rise to severe digestive upsets and damage to the nervous system, although the latter is uncommon. There is a strong possibility that exposure to arsenic compounds can give rise to cancer of the lungs, the lymph glands, and the skin. Because of the relative toxicity of arsenic, the recommended permissible limit of concentration in the workplace has been kept very low, and has recently been as low as 0.2 milligrams per cubic meter of air.

LEAD

Lead is one of the most common industrial poisons, either in its metallic form and combinations or in the compound known as tetraethyl lead, which until recently was used as a gasoline additive.

Lead is generally absorbed into the body by ingestion or by inhalation of lead fume. Commonly the exposure occurs from the ingestion of uncontrolled lead dust, either through direct contact with the mouth or by transfer from the hands. When absorbed into the body, it accumulates in the bones and builds up there over a period of about 5 to 20 years, during which time relatively little effect is observed. When the bones become saturated, the lead spills over into the blood stream again and begins to exert widespread effects throughout the body, disturbing the function of all body organs and systems. The initial effects, which include weakness, lassitude, and a general feeling of sickliness, are vague, generalized, and easily mistaken for some other problem. When the condition becomes established, however, three other classes of effect can be defined.

Effect on digestive system

As is true of the other effects, these tend to be rather vague. They include conditions such as loss of appetite, indigestion, constipation, and griping, the latter being a form of abdominal colic which can be extremely severe and even lead to hospitalization.

Effect on blood system

When it gets into the blood, lead passes into and begins to destroy the red blood cells. These cells are used for carrying oxygen to the tissues. As more and more cells are destroyed, the effects become manifest as *anemia,* which contributes to the general weakness and lassitude, and if allowed to continue will ultimately contribute to the resulting incapacity and even death.

Effect on the nervous system

One of the earliest effects on the nervous system is found in damage to the nerve supply of the hands and feet. Muscle weakness at the wrist and ankle, in fact, may be one of the first things other than the general unwellness that draws the attention of the victim to his/her condition. It occurs as a dragging of the foot (foot drop) or loss of ability to hold the wrist in the horizontal or raised position (wrist drop). There may also be shaking and trembling, and ultimately full paralysis.

The most serious effects, however, occur when the lead gets into the brain. This is very uncommon, although there have been reported cases with a curious condition called *pica.* The condition occurs among young children living in poverty circumstances who may be found perhaps chewing on wooden window sills covered with old lead paint, which they appear to find tasty because of its tangy flavor. Tetraethyl lead also exerts its major effects on the brain. When it was used as a gasoline additive, stringent measures were taken during its manufacture to ensure that no exposure was permitted. It is possible also that tetraethyl lead may have affected children living close to well-travelled roadways with a high exposure to motor vehicle exhaust.

The hazard from lead is very high, particularly in conditions where the general standard of hygiene is low and lead dust is allowed to accumulate or lead fume is emitted uncontrolled. The recommended permissible limit for concentration of lead in the workplace for lead dust and fume is 0.15 milligrams per cubic meter of air.

MERCURY

Mercury is in widespread use in the manufacture of electrical equipment and instrumentation as well as in the production of agricultural and industrial poisons, pharmaceuticals, and in the manufacture of measuring instruments containing mercury such as thermometers or barometers. As a shiny liquid metal, it was held by the medieval alchemists to have almost magical properties, and indeed, the great Paracelsus in the 16th century was the first to use it in the treatment of syphilis, which occurred as an epidemic in his time. His treatment, in fact, was essentially still used right into the 20th Century, although many patients were injured almost as much by the treatment as by the syphilis.

Mercury, as we have seen, can give rise to severe brain and nerve damage. However, it causes a lot of other problems as well. In particular, the fume is an irritant to the air passages and mucous membranes, and after the mercury gets into the body, it can cause digestive disturbances, excessive production of saliva, loss of appetite, loss of weight, fatigue, headache, sleeplessness, and irritability, as well as psychological and nerve disturbances. On top of that, it has a cumulative effect because it is not rapidly excreted. As a result, it tends to be deposited in the liver and kidney in particular and results in damage to these organs.

Mercury has an even greater hazard potential than metallic lead, and because of that the limiting value has recently been maintained at 0.05 milligrams per cubic meter of air.

METAL FUME FEVER PRODUCERS

The word fume is often misused. As we have seen, it doesn't mean smoke, or vapor, or gas. Strictly speaking, fumes are very fine solid particles floating in the air that have been solidified out of the hot gassy state of a metal or a metal compound. For instance, if you grind zinc or weld brass, some of the metal is heated sufficiently to become a metallic gas. As that gas cools, it condenses into fine fumes that can then be breathed into the body.

Metal fume fever is the name given to a curious illness that can occur when some of these fumes are inhaled. It is also known as brass chills, foundry ague, or foundry shakes, among other names. The condition can occur in brass foundries, during operations in the galvanizing industry, during welding, or in processes where molten metals, particularly zinc, are used. The cause is not really understood, but the condition is well-defined. The metals involved include zinc in particular, as well as manganese, magnesium, and copper, with the copper commonly being in the form of copper alloys such as brass. In the next chapter we'll see the misfortune that befell Johnny Calabrese when he unwittingly found himself exposed to some fuming zinc.

CHEMICALS THAT DAMAGE THE BLOOD CELLS

The blood is made up of a liquid portion called the *plasma* and a solid portion which in turn comprises both *red cells* and *white cells.* The red cells, which are formed in the bone marrow and then passed into the blood stream, contain a chemical called *hemoglobin* that has the capacity to carry oxygen to the body tissues and carbon dioxide away from the body tissues. The white cells are involved in maintaining immunity to infection and in fighting disease. It is the interference by a virus with the immume process of the white cells that give rise to what we know as AIDS (auto immune deficiency syndrome). This condition, however, does not occur from exposure to chemicals.

A number of chemicals, however, besides lead, which we have already looked at, have the capacity to damage the constituents of the blood and the bone marrow where the blood cells are made. Some act to destroy the red and white cells themselves; some act to block the hemoglobin and prevent it from carrying oxygen; still others actually change the chemical nature of the hemoglobin, while some act directly to irritate and eventually destroy the bone marrow. A few of the more common are considered below.

CARBON MONOXIDE

By far the most common of the chemicals that act on the blood is carbon monoxide. Carbon monoxide is a colorless, odorless, and tasteless gas that can occur whenever there is low-level burning. Consequently, potentially dangerous exposure can be found as a result of arc or gas welding, unventilated internal combustion engines, as well as in foundries, in closed rooms with fueled stoves or heaters, or both. It can even be found in the blood of tobacco smokers. Indeed, the heavy smoker who smokes a pack or two a day has enough carbon monoxide in his blood stream to reduce the oxygen level in his blood to the level you might expect when breathing air at about 5,000 feet. And, of course, any additional carbon monoxide from some industrial process will reduce his oxygen level still further.

Carbon monoxide acts by blocking the capacity of the blood hemoglobin to carry oxygen. Hemoglobin doesn't actually combine with oxygen the way, for example, iron combines with oxygen to form iron oxide, or rust. Instead, each molecule of hemoglobin gloms on to the oxygen in a manner that you might fancifully consider a lobster would carry something in its claws. The process is called *chelation* (Greek, *chele:* a claw). And it won't let go. Unfortunately for us, the hemoglobin is more attracted to carbon monoxide than to oxygen in the same circumstances. Consequently, in the presence of carbon monoxide, more and more hemoglobin gets blocked and less and less oxygen is carried.

As we saw with Mrs. Madison, the effects of carbon monoxide at relatively low atmospheric concentrations, such as 50 to 100 ppm, can be serious but are insidious in their onset so that the victim may not be aware of trouble until it is too late. These effects include headache, which may become severe and prolonged, nausea, dizziness, and weakness. As the condition develops, there will be mental confusion and ultimately unconsciousness, perhaps with convulsions. At lesser concentrations, of course, the effects will be less severe, while at high concentrations (for example, 0.3% to 0.4%) the effects can be dramatic, with rapid unconsciousness and death.

ANILINE AND ITS DERIVATIVES

Aniline is a purple dye that has the distinction of being the first synthetic dye ever made. It was developed in Germany at the end of the 19th century and its manufacture led to the development of the entire synthetic chemical and pharmaceutical industry that we know today. It is used as a dye and also as a stage in the synthesis of other dyes and chemicals. Unfortunately, both aniline and its derivatives, such as monomethylanaline and dimethylaniline, are toxic.

Aniline acts by combining with the hemoglobin in the bloodstream to form a new compound that cannot carry oxygen. This mechanism is not the same as with carbon monoxide. In the case of aniline, there is not merely a blocking of the capacity of hemoglobin to carry oxygen, but an actual formation of a new chemical called *met*hemoglobin. The effects are pretty much the same however, although the effects of carbon monoxide are much more dramatic and rapid in onset than those of aniline and other methemoglobin formers. There can also be one rather striking difference. When there is a relatively high concentration of carbon monoxide, the lips and skin of the victim tend to turn a bright cherry red color because of the color of the carbon monoxide hemoglobin. On the other hand, with ani-line and its derivatives, the lips and skin tend to turn a dusky blue in a con-dition that is known as *cyanosis,* which, in fact, may be the first evidence of any poisoning and may be noted by a victim's fellow employees. As with carbon monoxide, headache, dizziness, and weakness also occur, although the more dramatic effects of carbon monoxide poisoning do not tend to occur because the condition is normally discovered long before that level of damage takes place.

Because a higher incidence of cancer has been found among workers in the dye industry, at one time exposure to aniline was thought to be a cause of cancer. There has, however, been no evidence to substantiate this assumption. It is now believed that any increase in the incidence of can-cer among aniline workers is associated with the presence of other chem-icals rather than with aniline.

BENZENE AND ITS DERIVATIVES

Note the spelling. Benz*i*ne, with an *i,* is an old name for gasoline that is still used in some parts of Europe. Benz*e*ne, with an *e,* is a synthetic chem-ical that is used in the production of other chemicals, as well as being a solvent in its own right.

We have already seen, when looking at some of the effects of solvents, that benzene is an irritant and also acts to depress the function of the cen-tral nervous system. By far its most serious problem, however, is its action on the bone marrow.

Most of the bone marrow is found within the center pulp of the long bones and the breast bone. Bone marrow manufactures red blood cells and some white blood cells. Long-term, low-level exposure to benzene acts to initially stimulate, and eventually destroy, this activity.

Thus, while effects on the CNS will tend to occur on exposure to con-centrations in the range of 200–500 ppm, exposure to levels around 50–100 ppm may not cause depression of brain function, but can act to dam-age the bone marrow and interfere with its function of producing blood cells. This damage can give rise over a period of years to the general pic-ture of *anemia,* with weakness, lightheadedness, headache, nausea, loss of appetite, and perhaps even breathlessness, as the number of red cells in the blood becomes inadequate to support the oxygen-carrying activity. The victim will probably be pale and will tend to bleed easily, even sponta-neously bleeding at the gums or nose. Eventually, as more and more of the bone marrow is destroyed, the simple anemia may develop into what is called *aplastic anemia,* in which the bone marrow is no longer effective. Ultimately, unconsciousness and even death can occur.

Some authorities suggest that still another condition can be found in some persons. This is a form of cancer of the white blood cells known as *leukemia.* Leukemia, which can occur in various forms depending on the cells involved, is an excessive overdevelopment of white blood cells. It is as though, while preventing the manufacture of red cells, the benzene in the bone marrow actually stimulates the manufacture of white cells to grossly excessive amounts, which will ultimately kill the person exposed.

CHEMICALS THAT CAUSE SENSITIZATION

Several chemicals, in addition to being intensely irritant, have the capacity, particularly in susceptible persons, to provoke what is known as *sensitization.* Sensitization is a peculiar disorder of the so-called *immune system* of the body. The immune system is a widespread complex of cells and chemicals that is responsible for developing and maintaining immunity to infections and assaults from foreign substances that get into the tissues. Normally, when an infection occurs, the immune system goes to work to contain that infection and destroy the invader. Sometimes, for reasons that are only partially understood, those actions get out of hand, and the affected body tissues overreact, sometimes in a violent fashion, to what may be an exposure to very small or even minute quantities of the sensitizing agent. Sensitization doesn't happen to everyone, only to those who for some reason are peculiarly susceptible, and it may take varying periods before it develops. In some persons, sensitization might take several exposures over a few days; in others, it might take many exposures over months or even years.

One of the manifestations of this problem is a condition called *bronchial asthma.* Bronchial asthma is caused by sudden contraction, or spasm, of the bronchial breathing tubes that lead to the lungs. It can have many causes, most of which have nothing to do with industrial exposure. In industrial cases, the sensitizing agent, which is normally a specific chemical, acts to sensitize the tissues in such a manner that any subsequent exposure will produce an attack of asthma. Sensitization can occur very suddenly. A worker may be exposed to the agent on a regular basis or intermittently for some months with little more effect than, for example, irritation of the eyes and the breathing tract, and then suddenly, with no change in the exposure, he or she may almost instantaneously go into a violent asthmatic attack. Once the sensitization has developed, the potential is present permanently, so that any further exposure will bring on another attack.

The attack itself can be violent and disabling, with coughing, spitting, wheezing, difficulty in breathing (particularly in exhaling), and gross discomfort. It can last for minutes to hours. It may diminish spontaneously or it may require active medical intervention. Continued repetition can cause permanent damage.

While asthma is a dramatic and potentially dangerous form of a sensitization reaction, another less dangerous but also disabling form can occur when the sensitized tissue is the skin. In this case, the condition takes the form of allergic or sensitization dermatitis, which is manifest by a reddened and inflamed reaction in that area of the skin exposed to the sensitizing agent. Dermatitis is a general name given to any inflammation of the skin. *Allergic dermatitis,* however, is a sensitization reaction that can occur dramatically and suddenly after exposure to a chemical to which the skin has previously become sensitized. Almost any irritating chemical can cause sensitization in a given individual if he or she is susceptible, although some are more provocative than others.

Fortunately, as far as asthma is concerned, the significant agents are relatively few in number and not commonly encountered except in specialized industries. Table 7-3 lists some of those concerned.

Table 7-3
Chemicals That Can Cause Sensitization

Name	Usage	Effects	Hazard Potential and PEL
Methyl diisocyanate (MDI)	Making polyurethane foam	*Irritation:* eyes, mucous membranes, breathing *Sensitization:* asthma	Very high (0.02 ppm)
Methyl isocyanate	Making polyurethane foam	*Irritation:* eyes, mucous, membranes, breathing *Sensitizaton:* asthma	Very high (0.02 ppm)
Phenylenediamine (Ursol)	Making furs, rubber, dyes	*Irritation:* skin, breathing, mucous membranes *Sensitization:* skin	High (TLV 0.1 mg per cu. meter)

(continued on next page)

Table 7-3 (continued)
Chemicals That Can Cause Sensitization

Name	Usage	Effects	Hazard Potential and PEL
Toluene-2, 4-diisocyanate (TDI)	Making polyurethane foam	*Irritation:* eyes, mucous membranes, skin Sensitization: Asthma	Very high (0.005 ppm)
Wood dust (particularly red cedar)	Many	*Irritation:* Eyes, mucous membranes, *Sensitizaton:* Asthma	Moderate (N/A)
Grain dust	Grain processing	Irritation: eyes, mucous membranes, *Sensitization:* Asthma	Moderate (N/A)

Sources: multiple

OTHER CHEMICALS

This chapter covered only a small proportion of the vast number of chemicals that can be found in the workplace. For more definitive information on a wide variety of toxic substances, the reader is referred to standard references on industrial toxicology. Of these, one of the most useful for a quick summary of toxic effects and management is *Chemical Hazards in the Workplace* by N. H. Proctor and J. P. Hughes. Definitive reviews of many chemicals are published by the American Industrial Hygiene Association and the National Safety Council among others. Sources are given in Appendix III.

DUSTS

Dust in the workplace is often considered to be merely a nuisance that has to be cleaned up once in a while. In some cases that's true. In others, however, as we have seen with lead, dust can be a significant source of hazard. The problems associated with dust will be considered in Chapter 9.

The Strange Case of Johnny Calabrese

Johnny was a welding apprentice who went part-time to technical school and part-time to his job in a metal fabricating plant. On this particular day, he was given the job of brazing together pieces of galvanized zinc evestroughing. It wasn't a very difficult job, and he worked happily at it all that day. In fact, while he was working, he spent most of the day thinking about the new girlfriend he was going to spend the evening with. So when he met her at 6 o'clock that evening, he was all set for a big time. However, fate and some galvanized zinc decreed otherwise. By 7 o'clock he was beginning to feel unwell, in spite of his expectations, and by 7:30, he was cold and shivery and interested in nothing more than crawling into bed, with a couple of aspirins—which he duly did. By the next morning, he felt pretty good again, good enough in fact to go back to work. He returned to the same job, completed another day's work and set out once more to meet Sarah. But he must have felt the fates were against him, because the same thing happened again. More chills! More sweats! More miseries! Stomach ache and nausea! Sarah went back with him this time, and spent most of the night looking after him, trying to calm his shakes, ease his thirst, and, to put it in romantic terms, soothe his fevered brow. Once again, Johnny was better by morning, although he stayed home from work this time. He returned to the same job on the following day, not recognizing, of course, that the job was causing the problem. On this occasion, however, the attack was much less severe, just a little bit of discomfort that didn't prevent him from dating again that evening.

So what was it all about? Johnny Calabrese had, in fact, suffered a relatively minor attack of metal fume fever. He represents the classic case of a worker spending the day exposed to the hazard of inhaling certain metal fumes. After completing his day's work, he goes home feeling tired but otherwise well. Some hours later, however, perhaps as much as eight or as little as four, he begins to feel miserable with chills, shakes, and generalized discomfort, perhaps even with nausea and vomiting. If he were to take his temperature, he would find he had a fever of several degrees above normal, and would think he was coming down with the flu. However, by next morning, or certainly within 12–24 hours, he would be completely recovered—until the next time. Eventually, after a few recurrences, he would develop an immunity that would last as long as the exposure was repeated, although the immunity would tend to be lost within two to three days away from exposure. This condition is often experienced by a new employee on his/her first day's work or a veteran employee who returns after some absence.

CHAPTER NINE

How Deadly Is Dust?

DUSTS

Dust in the workplace is often considered a mere nuisance that has to be cleaned up once in a while. In some cases, however, as we have seen with lead, dust can be a significant source of hazard.

Three classes of dust can exist in the workplace, namely, *toxic* dusts, *inert* dusts, and *proliferative* dusts. We have already seen the effects of some of the toxic dusts, such as lead. The other two groups are considered below.

INERT DUSTS

So-called inert dusts, which can be distinguished as chemicals, are deemed to be relatively harmless, and, in general, can be permitted to exist in the breathing air up to a level of about 10 milligrams per cubic meter. As we shall see, however, some of them can cause trouble if breathed in large quantities. The inert dusts include materials such as carbon (smoke or soot), calcium (limestone dust), iron oxide from welding, grinding, cutting, and burning of iron or steel, abrasives of various sorts such as carborundum, emery, and aluminum oxide, mica from electrical equipment and the manufacture of wallpaper and rubber, and kaolin from china clay.

These materials get into the upper air passages, and indeed can be irritant. Over prolonged periods of time, they can also accumulate in the lungs. If you were to examine the lungs of a smoker, for example, or someone like a welder, and if he or she were repeatedly exposed to smoke without protection, you would find after a few years that his or her lungs were black with carbon. In someone working with iron dust, you might even be able to see it on X-ray. The effects of these dusts, however, are largely mechanical. They may give rise to a chronic cough with spitting, shortness of breath, and a greater likelihood of developing lung infections such as bronchitis or pneumonia, but, unlike the proliferative dusts, they do not actually destroy the lungs. In fact, unless the accumulated volume is excessive from long-continued exposure to very high levels, there is very little measurable effect on the lungs at all, provided that the person concerned has not been unfortunate enough to develop an accompanying bronchitis. Breathing performance is evaluated by lung function testing, in which various characteristics of breathing are measured by specialized instruments. The presence of inert dusts rarely causes measurable disturbance of lung function.

The accumulation of carbon, of course, is not the cause of the lung cancer found in heavy smokers. That occurs from the presence of other chemicals.

There is one special group of inert dusts. That group makes up the *biological* dusts. Biological dusts are derived from living, or recently living, materials such as wood, grain, flour, starch, and sugar. Most of them are relatively harmless, although on prolonged exposure they can give rise to irritation of the breathing tract and the skin. Sugar is particularly irritating to the skin. In some susceptible persons, allergies, such as asthmatic reactions or allergic dermatitis can occur.

Cotton dust presents a special case. In days gone by, not so long ago, when the ventilation in cotton processing or clothing plants was next to useless, the breathing air was so thick that workers couldn't see across the room. Cotton dust would block the nose and the sinuses, and of course, accumulate in the lungs. In susceptible persons, a disabling asthma-like condition could occur, characterized by shortness of breath, weakness, and recurrent attacks of coughing, wheezing, and tightness of the chest, eventually leading to total disability and even death. Fortunately now, because of improved working conditions, the occurrence is relatively rare, although some of the victims are still around.

PROLIFERATIVE DUSTS

The word *proliferate* means to grow or expand. Proliferative dusts accumulate in the lungs where they destroy the functioning lung tissues and replace them with fibrous scars, along with nodules or growths that tend to expand and grow. As the lung tissue is destroyed, and as the growths develop, the functional capacity of the lung becomes more and more decreased until, over a period of years, it can no longer support life. In addition, there may be an accompanying bronchitis, or worse still, infection from the once deadly disease known as tuberculosis. Two materials, in particular, are responsible for the production of this condition. One of these is silica in its various forms, and the other asbestos.

Silica

Silica, or silicon dioxide, occurs in various forms including chalcedony, which is a decorative material; chert, which is used in abrasives; flint, which is used in abrasives and ceramics; jasper, which is used for decorative purposes; quartz, which is a constituent of sand; tripoli, which is found in scouring powders, polishers, and fillers; cristobalite, which is used in high temperature casting and specialty ceramics; diatomaceous earth, which is used in filtration processes and as a filler; and finally, silica gel, which is used in dehydrating and drying. Note, however, that the material of concern is *silica,* and not *silicates,* which are relatively harmless derivatives of silica, nor *silicones,* synthetic materials used especially as lubricants. Neither silicates nor silicones cause proliferative conditions.

Exposure to silica, as you might imagine, is considered to have a high hazard potential. The recommended limiting value for cristobalite as a total mass in the breathing air, for example, is 0.15 milligrams per cubic meter of air.

The condition that can occur from exposure is known as *silicosis.* It is characterized, as are the other proliferative disorders, by coughing, shortness of breath, and progressive loss of lung function.

Asbestos

Until recently the potential for exposure to asbestos was widespread. It occurs, of course, in mining and milling of the material, and in processing it into industrial and consumer goods. Asbestos, however, is also sprayed on walls, ceilings, pipes, and cables for thermal and electrical insulation in public and office buildings, in plants and factories, as well as in residencies and ships. It occurs in fire smothering blankets and safety clothing, as a filler in plastics and cement, in pipes, and in asbestos tile.

Recently, because of widespread concern, major attempts have been made to stop its continued use and to remove it from those locations where it has already been in existence. Unfortunately, some of these removal attempts have not been as carefully undertaken as perhaps they should, and free asbestos in some instances has been disseminated into the breathing air.

Like silicosis, the condition of *asbestosis* gives rise to a progressively increasing cough, difficulty in breathing, and loss of lung function. In addition, however, asbestos has also been found to cause cancer of the lung as well as a rare form of cancer of the membrane that surrounds the lung and the membrane that surrounds the abdominal organs. Like silica, asbestos is considered to have a high hazard potential.

Control of exposure to dust and other chemicals will be discussed when we examine the management of toxic exposure. But let's look for a moment at a situation where there wasn't very much control over any exposure, and where, even if there had been, it wouldn't have made an awful lot of difference to our next hero, poor Jake Whitman.

Jake Whitman's Plight

Jake Whitman had a problem. It was a problem he didn't want to think about, and certainly one he didn't want to talk about. But nevertheless he did think about it, off and on each day, as it crossed his mind or something happened to remind him, until, as the days passed into weeks, he was finally forced to admit the truth to himself. He was impotent, although he didn't call it that. He didn't even know the word.

He hadn't always been impotent. Far from it, as three kids and a happy marriage to Maria could attest to. And at age 38 he certainly couldn't call himself senile. But recently, over the past seven or eight weeks, it had become more and more obvious that, as he put it to himself, he could no longer make it.

And so Jake was worried. He felt vulnerable in some indefinable way, as though somehow his very manhood was being threatened, which perhaps it was, although he couldn't have said that in so many words either. He had never felt insecure before, never lost confidence in his ability to earn his living, nor even to tackle with his fists, if necessary, anybody who was foolish enough to challenge him. For Jake was a big man, a friendly giant, strong and powerful, the last man you would think could ever become impotent. And yet here he was, he couldn't even say it.

Now Jake wasn't very bright, but he'd heard the guys talk, and he'd laughed with them at the jokes told over a few beers when they went out together. He knew that this could happen to any guy once in a while or perhaps even more often when something wasn't quite right, or when he was sick, or drunk, or something. But him! All the time! Day after day! Weeks! There had to be something wrong. And yet he wasn't sick, except for the occasional stomach ache and indigestion that didn't amount to much.

So Jake worried and worried, and the more he worried the more distracted he became. He began to lose his appetite and to lose weight. He couldn't sleep at night, he didn't seem to have the strength he used to, and his indigestion and stomach ache got worse He began to get into trouble at work because he was taking time off, and his work wasn't up to par even when he was there. Finally, pushed by Maria, he went to see the doctor.

Dr. Gates was just recently qualified. He was a personable young man, knowledgeable, and up-to-date with all the latest tests and treatments. But, like many doctors, however well-intentioned, he didn't know much about occupational medicine, or industrial toxicology, or even what went on behind the factory fence. For that matter, he had never even been behind a factory fence, far less inside the factory itself. And throughout the whole of his arduous training in medicine he had never encountered anything but passing reference to any medical problems that might derive from work or the workplace. He had never received any specific training in occupational medicine, and the only reference to toxicology he had met was in some aspects of his course in forensic pathology, which is the branch of medicine that deals with crime, criminals, and police autopsies. But he was a good doctor, and the mission to which he had genuinely dedicated himself was to help, to the best of his ability, those who were sick or injured.

So perhaps we can't really blame him for his reactions when Jake came in, sat down beside his desk, and said bluntly:

"Doc, I can't . . . you know . . . I can't . . . make it . . . with Maria."

"Well, that's not too much of a problem," said Gates, in his best professional manner. "How long's that being going on?"

So Jake told him, somewhat reluctantly, about his worries over the past weeks, and about Maria, and about his three kids, and about his fears.

"You see, it wasn't always like that, Doc," he said. "I mean, Maria and I had it good. And before Maria, well, I wasn't a young stud, you know, but I was OK. But now . . ." He stopped, looked down, shuffled his feet, and went on, embarrassed. "I still like to look at her, you know, and . . . touch her, and . . . things. But, you know, nothing happens."

Now, as I said, Dr. Gates was a good doctor, and he knew that most impotence originates in the head rather than in the tail. But at the same time, since it seemed to have occurred so suddenly, he thought there must be something else there, perhaps some physical cause. So he started making inquiries, and before too long he learned about the sleeplessness, the indigestion, and the stomach ache. He asked about Jake's past history, and his family history, and he even asked what kind of work he did, although it didn't really help much when Jake explained he worked in the wire making plant where he tended the quenching bath. He did a physical examination, and some routine urine tests, and even some routine blood tests. He arranged for special abdomenal x-rays. He checked him over for cancer, diabetes, anemia, and a few obscure diseases that might by some stretch of the imagination cause his problems. He even wondered if he could have "yuppie flu," as they were calling it these days. But Jake was certainly no yuppie. In fact, other than a mild anemia, Dr. Gates couldn't find anything the matter. So when Jake came back to his office, all that the good doctor could say was:

"Well, Jake, you seem to be as fit and healthy as anyone else. I can't find anything the matter with you physically. But you know, something like this often occurs just from stress, and certainly it is aggravated by worry. So maybe I should send you to see a psychiatrist."

But Jake wasn't "goin' to see no shrink," as he put it, so he went back home with some iron pills for his blood, indigestion pills for his stomach, and sleeping pills for his bad nights.

Things settled down a bit over the next few weeks. His stomach ache and indigestion diminished a little with the medicine, and he could sleep at nights with his pills, even though his love life was no better off than before.

And then something happened that changed his life. The plant was fairly small, with relatively few employees. The wire was produced from metal rod stock, which was heat-treated in a furnace, quenched by drawing it through a long trough of lead, descaled and further processed before being passed to the drawing machines. Employees were rotated to a certain extent through various jobs, although Jake spent most of his time tending the quenching. Most of the employees, including Jake, were veterans who had been at the job for 10–20 years. Jake himself had started when he was 22, about the same time, in fact, as when he got married.

Anyway, something happened, and the something that happened was that the sickness absenteeism rate in the plant began to rise alarmingly. Shop floor employees would take sick leave with the same kind of complaints as Jake's—except for the impotence. Jake was unique in that, perhaps because he was so big that he didn't get as sick with the other symptoms, or perhaps merely because the other guys didn't mention it.

Now, the manager knew that in using a lead bath he was working with a potentially hazardous substance, and he knew there were regulations governing the use of lead, but he had a hood over the trough, and he had exhaust fans in the hood. And, most significantly from his point of view, he hadn't had any trouble up until now. There hadn't been any trouble up until now, of course, because when lead gets into the body it is isolated in the bones for some 10–20 years before it begins to spill out and cause trouble. So he hadn't been too concerned, and he'd never made any measurements or monitored his workers. But now, half his small workforce was in trouble, and he knew that he too could be in trouble with the law if he didn't do something quickly, or perhaps even if he did.

In approaching panic he called in the Occupational Medical Consultant and some hygienists. The hygienists measured the concentration of lead in the breathing air, and, of course, in spite of the hood and ventilation, which were inadequate, found it far above the recommended level of 0.15 milligrams per cubic meter. The medical specialist examined the employees in the light of his special knowledge and sent blood samples to the lab to determine the lead content. Sure enough, he found that most of the men, including Jake of course, had blood lead content well in excess of what would be considered permissible. Jake, in fact, was not the most heavily affected. It just so happens, sometimes, that one person may be primarily affected in one way, while another is affected differently. Impotence, however, perhaps because its occurrence is so overwhelming, can be one of the symptoms of lead exposure that is of most concern to the victim—even if at times he doesn't want to talk about it.

I won't go into the details of treatment, procedures, time off from work, and structural modifications in the plant. Suffice it to say that about six months later, on Maria's birthday, Jake and Maria had a very successful celebration.

CHAPTER ELEVEN

Basic Toxicity and How to Measure It

We have seen that many different factors can contribute to chemical hazard in the workplace. The degree of hazard, however, is fundamentally determined by two factors: the basic toxicity of the agent concerned, that is, its intrinsic capacity to damage or affect biological tissue; and the severity of the exposure, or what is sometimes called the dose-response relationship. The duration of the exposure, of course, must also be considered.

BASIC TOXICITY

Basic toxicity has been identified from careful observation and experimentation in the workplace and in the lab. Over the years, from the time of the ancient Greeks and Romans, and probably long before that, we have learned that exposure to certain substances can cause bodily harm. Hippocrates, the founder of medicine in Ancient Greece, described the occurrence of lead poisoning among lead miners and metal workers as long ago as 400 B.C. The Roman historian, Pliny, described in his encyclopedia in the second half of the first century A.D., the dangers of mercury poisoning from the dust of mercury ore, and even devised a dust mask made from an animal bladder to protect the workers. Most of that work was forgotten, however, for 1,500 years, but eventually, as more and more knowledge was obtained, we became more and more aware of potential dangers. But it wasn't until this century that doctors, scientists, and others began to collect and organize the body of knowledge that came to be known as industrial toxicology. With that knowledge, derived from sickness records, absenteeism records, personal experiences, in-plant observations and measurements, as well as from laboratory experimentation on human volunteers and on animals, we are now in a position to define with reasonable certainty the hazards associated with exposure to virtually all the common industrial chemicals and many of the more unusual. Some of these have been discussed in Chapter 7.

It should be obvious, however, that the nature of the response to a given chemical hazard must also depend on the concentration of that chemical in the environment.

THE DOSE/RESPONSE RELATIONSHIP

As you might imagine, the greater the absorbed dose or the higher the exposed concentration and the longer the duration of the exposure, the more rapid and/or more serious are the effects. This relationship, however, is not what is called linear. In other words, for example, an increase in concentration at the lower end of the scale does not produce the same increase in the effects as it would in the middle range of the scale. It is only in the large mid-range of a possible concentration that there is a direct relationship between the amount of the dose and the extent of the response. At the lower end of the concentration scale, for example, a moderate change can elicit a comparable response, while at the upper end, perhaps close to or at the concentration that would cause death, an increase in concentration makes little or no difference.

One measure of dose/response that you might find quoted in the literature is what is called the LD_{50}. This is the median (average) oral *lethal dose* in terms of weight of substance per kilogram of subject and refers to the dose at which 50 percent of a population of the same species (say, mice or man) would be killed. When administration is by inhalation, the LD_{50} is termed the LC_{50}, or *lethal concentration,* in terms of parts per million (ppm) in air that would kill 50 percent of a population. Examination of the LD_{50} or LC_{50} of various chemicals can provide some information on their relative toxicity, keeping in mind, of course, that the respective data are derived under controlled conditions that do not necessarily apply in an industrial setting.

The relative toxicity in terms of lethal dose or concentration is shown in Table 11-1.

Table 11-1
Classes of Toxicity as Determined in Rats

Rating	Description	LD_{50} (oral)	LC_{50} (inhalation)
1	Extremely toxic	1 mg or less	less than 10 ppm
2	Highly toxic	1–50 mg	10–100 ppm
3	Moderately toxic	50–500 mg	100–1000 ppm
4	Slightly toxic	0.5–5 g	10,000–100,000 ppm
5	Practically non-toxic	5–15 g	10,000–100,000 ppm
6	Relatively harmless	more than 15 g	more than 100,000 ppm

Source: Hodge, H. C., and Sterner, J. H. Tabulation of Toxicity Classes. American Industrial Hygiene Association Quarterly, Vol. 10, 1949, p. 94.

Another way of defining toxicity is in terms of the amount of chemical it would take to produce death when swallowed by an average (150 pound) man, as shown in Table 11-2.

Table 11-2
Amount of Chemical Required to Produce Death When Ingested by a 150-pound Man

Rating	Description
1	Practically non-toxic, takes more than one quart (2 lb)
2	Slightly toxic, takes one pint to one quart
3	Moderately toxic, takes one ounce (30 g) to one pint
4	Very toxic, takes one teaspoon (4 ml) to one ounce (30 g)
5	Extremely toxic, takes 7 drops to one teaspoon 4 (ml)

Source: Derived from text in Hodge, H. C. and Sterner, J. H. Tabulation of Toxicity Classes. American Industrial Hygiene Association Quarterly, Vol. 10, 1949, p. 93.

There are still other factors, however, that can affect the response. One of these is the rate of absorption of a chemical. The faster the rate, the quicker the reaction. The rate of absorption in turn is affected by the way a chemical is administered. The more direct the route, the quicker the absorption. Thus, for example, although obviously it doesn't happen in industry, the quickest absorption occurs when a chemical is administered directly into the bloodstream by a syringe.

In industry, the quickest absorption occurs when a chemical is inhaled as a gas, a mist, or a dust. The next quickest is by mouth, and the slowest is through the skin, and, of course, only for those fat soluble chemicals that can be absorbed through the skin. Most chemicals are soluble only in water, if at all, and cannot be absorbed through the skin. As we have already seen, the skin acts as a barrier to absorption for chemicals that are not soluble in fat or oil. For skin absorption, the larger the area of skin involved and the longer the chemical is in contact, the greater is the response.

CONCENTRATION IN THE WORKING ENVIRONMENT

Of all the factors involved, however, clearly the most important is the concentration of the agent or agents concerned in the working environment. The concentration, of course, refers to the amount of toxic material in the environment to which a person is exposed. Now, there may be 1,000 lbs, or even 1,000 kg, of a chemical somewhere in the plant. The total quantity is irrelevant for our purposes. What we need to know is how that chemical is distributed in the working environment, and in particular, how it is distributed in the immediate environment of the workers we are concerned about. If it is all in one place in a properly secured container, it will still be a hazardous material, but it will present little or no risk and certainly no danger. But if it is in the atmosphere as a gas or a vapor, then you will want to know how much gas or vapor is contaminating the air you are breathing. Obviously you can't measure all the air and all the gas or vapor, but you can take a representative sample of air and measure the proportion of gas or vapor found in that sample. The unit that is largely used for this purpose is the unit we have seen before, namely, *parts per million (ppm)*. This means that in a million volumes of air—perhaps a million pints, perhaps a million liters, perhaps a million milliliters—there will a certain number of similar volumes (or parts) of gas or vapor. If, for example, your instruments found a concentration of 400 ppm carbon monoxide in the air, then that would mean that for every million volumes of air there would be 400 volumes of carbon monoxide distributed in that air. Now that probably wouldn't be enough to kill you in the short term, but you would be feeling pretty miserable with severe headache, nausea, weakness, dizziness, and shortness of breath. You certainly wouldn't last more than an hour or so.

This practice of defining a concentration in parts per million is very common and useful, but it is only applicable to gases and vapors. When you deal with solids and liquids another measure is used, in this case *milligrams per cubic meter,* or the weight of the substance in milligrams as distributed in a cubic meter of atmospheric air. There are about 28 grams to the ounce, 1,000 milligrams to the gram, and about 35 cubic feet to the cubic meter. One milligram per cubic meter is a very small amount. Nevertheless, some compounds are so toxic that concentrations of even less than 1 milligram per cubic meter are necessary to ensure safety. The permissible concentration of lead fumes, for example, is 0.15 milligrams per cubic meter. On the other hand, for a less toxic material such as the solvent MEK (methyl ethyl ketone), the permissible value is 590 milligrams per cubic meter. Unfortunately, however, as has been pointed out earlier, permissible levels have not been determined for all chemicals.

If you realize that during moderately heavy work over an eight-hour day, you will inhale some 10 liters of air per minute, or 4,800 liters per day, this air will occupy a space of 4.8 cubic meters. Thus, if you know the concentration of a toxic material in milligrams per cubic meter, you can calculate how much of that material you will inhale during a day's work. For example, if the concentration of lead fume in the breathing air is 0.15 milligrams per cubic meter, normally the maximum acceptable, then in the course of a normal working day if you perform moderately heavy work you will inhale 0.72 milligrams of lead. That is not an intolerable load, but in the course of a lifetime of work, it does add up.

DURATION OF EXPOSURE

Still another factor to be considered is the duration of exposure. Although you may know that a certain material is toxic when inhaled and that the concentration of that material in the atmosphere is high enough to cause problems, you also have to realize that it is not going to cause these problems at that concentration unless you inhale it for a certain length of time. In industrial usage, we normally have to consider what the effects are going to be when that chemical is inhaled at a particular concentration over the duration of a normal working day. Thus we have to be able to measure that concentration in some definitive way.

MEASUREMENT OF CONCENTRATION

While it is easy to measure the duration of an exposure, it is not as easy to measure the concentration of an agent. There are various devices available for this purpose. Some can be used with little or no training, while others require the skills and services of a trained industrial hygienist or industrial hygiene technician. The greater the accuracy and reliability required, the greater the need for sophisticated instrumentation and the skill to use it. Some of the common devices are:

- dosimeter badges
- detector tubes
- cassettes and filters
- sampling pumps
- direct reading instruments

Each of these is considered below.

DOSIMETER BADGES

The dosimeter badge, or diffusional monitor, such as that made by the 3M Company, can be used by someone with no specialized training. It looks like a small open plastic case, about the size of a pocket watch, and is attached to the worker's clothing as near to the nose and mouth as is feasible. Inside the case is a material that can be penetrated by the outside air. The material contains a chemical that is specific to the contaminant or group of contaminants you want to measure. The contaminant becomes attached to the material during the course of the working day. At the end of the day, the badge is removed and sent to a laboratory where the amount of contaminant that has been absorbed is measured. From that measurement, the concentration of the contaminant in the atmosphere can be calculated. Many different dosimeter badges are available, for different purposes.

DETECTOR TUBES

Detector tubes, such as those made by the Draeger Organization or the National Mine Service Company, can be used for screening purposes by someone with minimal training, although it is generally wiser to have the work done by an expert. Detector tubes are pencil-shaped glass tubes, pointed at each end. Well over 150 different detector tubes are available for measurement of different chemicals. A list of these is included as Appendix III. Each tube contains a measured quantity of a chemical reagent, specific for a given contaminant. When the material in the tube is exposed to the contaminant, it changes color. In most circumstances, the color change takes place over a varying length of the reagent in the tube, depending on the concentration of the contaminant in the air. The length of the color change in the tube is measured against a scale on the glass and can indicate the concentration of the contaminant in the air. With some chemicals and detector tubes, the intensity of the color change gives a qualitative reading rather than a semiquantitative one.

To use a short-term detector tube for measurement of instantaneous concentration of a contaminant, the pointed ends are broken off and one end of the tube is inserted into a special hand pump supplied with the kit. The pump is then used to pump a specified amount of air through the tube. The technique is relatively simple and effective, although the results may not as precise as with more sophisticated methods. Depending on the contaminants and the conditions of use and the skill, there can be as much as a 25 percent error in some situations.

Special tubes are available that can be used for long-term average concentration, rather than instantaneous.

CASSETTES, FILTERS, AND PUMPS

A common method used by industrial hygienists and technicians for measuring concentration employs a device known as a cassette. A cassette is a clear plastic cylinder about 1 inch in diameter and up to 1½ inches in height. It is open to the air at one end and contains a filter that may be made of specially treated paper, glass fiber, or other material such as cellulose acetate, depending on what contaminant or contaminants are being measured. It is closed at the other end and incorporates a small fitting that can be connected by rubber tubing to a pump that allows air to be drawn through the open end of the cassette and through the filter. A specific filter is chosen according to the chemical and physical properties of the substance or substances to be collected, which could be particles, fumes, vapors, and so on.

The sampling pumps are precision devices capable of continuous and consistent action at up to 2 liters per minute for as much as 8 hours without variation in performance. In other words, they will always pump at the same flow rate, even as the cassette becomes more and more loaded. Each pump is about 6 inches wide, 5 inches high, and 2 inches thick and weighs a couple of pounds. It is battery operated. In practice, a suitable filter is installed in the cassette and the cassette is attached to the workers's clothing near the mouth and nose in order to sample the breathing air. The pump is attached to a waist belt and connected to the cassette by a tube. The pump is operated during the normal work cycle for a period sufficient to make an adequate sample for later analysis, commonly 4 to 8 hours. When the sampling period is finished, the cassette is dismantled and the filter removed. The material collected on it is analyzed in the laboratory to determine the concentration in the air. Pumps are also used (although rarely these days) with collection devices other than cassettes, such as cyclones or impingers, which essentially are glass tubes for the collection of chemicals or dust.

DIRECT READING INSTRUMENTS

Sometimes, instead of collecting samples for later analysis, hygienists occasionally use measuring instruments that can give a direct indication of the presence of a contaminant and its concentration in the atmospheric air. Special instruments, for example, are available for the evaluation of specific gases such as carbon monoxide and oxides of nitrogen. There are even some complex and sophisticated instruments such as portable infrared gas analyzers and mass spectrometers that can, in appropriate circumstances, be used to identify the presence and concentration of materials in the atmosphere that are unknown to the investigator or whose presence can only be guessed at.

So we can indeed identify and measure the various toxic contaminants that we find in the industrial environment, and we can compare these findings with the established limits of permissible exposure that will be discussed later. But sometimes there are situations where none of these procedures seem to have been applied. Let's look at another true story, this time about someone who didn't know anything about toxic hazards, much less about concentrations of toxic chemicals, and certainly didn't recognize the potential hazard of his environment.

The Man in the Next Motel Room

When David Wharton drove into Watkins Glen in upper New York State late that afternoon, he wasn't looking for a race track, although he knew there was a famous one nearby, nor did he want to take a scrambling walk alongside the falls as they tumbled down the glen. He'd done that before. What he wanted after ten hours of hard driving was a pleasant motel, a good restaurant, and a comfortable bed. So when he saw the place up on the hill, in what seemed to be an ideal location, he pulled in and got himself a room. And as he sat outside in the sunshine, looking across the valley from the balcony, and sipping on a drink before dinner, he considered that he had chosen well.

David was a doctor, a specialist in occupational medicine, with particular interest in the toxicological problems of industry, and, in fact, he was on his way home from a conference on the East Coast where he had presented a scientific paper on some of his latest research. So it was a pleasing coincidence that as he sat taking his ease, he should be confronted with a case the like of which he had never encountered before.

The balcony on which he sat was common to each of the rooms in front of the motel, although there was a pretense of separation by planter pots and decorative fences. When the occupant of the room next door also came out on to the balcony it was very easy for Dr. Wharton to greet him in a friendly fashion. In fact, it would have been almost rude not to do so.

"Beautiful evening," said David.

"Yeah, it sure is!" said the man. "Nice to relax after a long drive."

From what David could see, the man was a pleasant looking person, perhaps about 40, with a friendly manner and casually dressed, although he seemed curiously tense, as though wound up like a tight spring.

"Care to join me?" said David, holding up his glass.

The man hesitated a moment.

"Ah, yes, thank you. That would be very nice."

He stepped over the decorative fence and sat down beside David at the balcony table.

"Hold on while I get the bottle and another glass," said David. "By the way, I'm David Wharton."

"Glad to meet you," said the man, holding out his hand. "Nick Serducci."

When David returned, Nick had relaxed somewhat, leaning back in the chair. David poured the drink and raised his glass.

"Cheers!" he said. "Welcome to the Overlook Motel, or whatever they call it."

Nick again hesitated, and then put his hand forward to pick up his drink. As he did so, it began to tremble, and as it approached the drink, the tremor became so great that he had to use his other hand to steady it before raising both hands with the glass to his mouth. He took a swallow, carefully put the glass down, and turned to David with a wry smile.

"Cheers!" he responded, looking a little quizzical, and then added, "I suppose you think I must be a raving drunk?"

David gave an embarrassed laugh. It was exactly what he was thinking.

"Well, I don't know!" he said, "I have to confess it crossed my mind. You've certainly got quite a shake!"

"Yeah, that's true! But you needn't worry. I assure you I'm not a drunk. Never have been. Although I do take an occasional beer or a whiskey every now and then. In fact, for several years I never took a thing. As a matter of fact, I've only just started again, but every time I go to pick up the glass I have this problem. It gets embarrassing!"

David looked sympathetic but didn't say a word. He sensed a story but he didn't want to pry. At the same time, Nick seemed more than willing to enlarge on his problems.

"You know," said David, a little diffidently, after a moment, "I don't want to stick my nose in where it doesn't belong, but are you telling me you are not an alcoholic? You mean this is some illness, some hereditary problem?" He laughed, again a little embarrassed. "Don't answer if you don't want to."

Nick laughed too, and raised his glass again, this time with less tremor.

"Oh no! I'm not an alcoholic. And this isn't a hereditary problem, nor even an illness in the ordinary sense of the word." He looked up, expectantly. "But if you have a minute, I can tell you a story."

David put down his glass and leaned back, wondering what he was getting into.

"Sure!" he said, thinking of the Ancient Mariner. "Let's hear it."

Nick looked across the valley into the distance, as if he were remembering something and trying to marshal his thoughts.

"Seven years ago," he began, "I was in real estate, doing quite well, but, you know, it's a chancy business. One day you make $15,000 and for the next two months you don't make anything at all, and then you make another $10,000 and so on. You never know where you are. And so when I learned about a business for sale, I thought I would try it. Only part-time at first, while I carried on with the real estate. Anyway, there was this guy who had a small manufacturing business for sale. He just recently bought it himself, but he had to move away suddenly so he was selling it cheap."

Nick paused and took a drink, with the same far away look in his eyes. His tremor was markedly less now, and he no longer seemed to be aware of it.

"So I bought the business," he went on. "It was simple work, really. The guy had a contract to make medical thermometers for a drug store chain. He made them in this converted workshop. The workshop wasn't very big, maybe about 30 feet by 18 feet, with a door at one end and a big window at the other. There were workbenches round the walls, and water and power, and places for storage of raw materials and for cartons of the product. Pretty primitive, really. But it was a moneymaker. And the longer you worked, the more money you made."

"But what was the work?" asked David, becoming interested.

"Oh, there was nothing to it. I had these calibrated glass tubes with scales etched on them. You know what medical thermometers look like, don't you?"

David nodded.

"Well, I'd boil up some mercury in a glass beaker on a Bunsen burner. Then, I'd hold one of the tubes up in front of me and pipette the proper amount of mercury from the beaker into the tube and seal the ends. As the mercury cooled, it created the proper vacuum inside. God, it was monotonous work! Hour after hour, day after day, for about two years. I didn't have a tremor then, of course. In fact, I needed a pretty steady hand not to waste the mercury. It was expensive stuff!

"Anyway, all went well for a year or so, and the money started pouring in. But then, after a while, things began to go funny. I began to have more and more difficulty getting the mercury into the tube. I wasn't really aware of a tremor, but I didn't seem to be able to coordinate my movements like I used to. And the more difficulty I got into, the angrier I would become. I'm divorced now, but my wife at that time used to complain that I became annoyed over nothing at all. I guess I used to get angry very easily, and I would rant and rave one minute and be all sweetness and light the next. Not like me at all. Or at least, not like I used to be. I used to be very even tempered. I suppose that's why my wife left me. She used to get frightened, and she couldn't figure me out at all. However, sometimes I would feel real good, higher than a kite, the world was great and everything was wonderful. And then the depressions would come. I couldn't make up my mind about things. Couldn't make what should have been simple decisions. Couldn't get my orders straight, and the customers would complain. Know what I mean?"

"Sure, I know," said David, now very much aware of what was going on. "Mercury poisoning. You lost weight, and couldn't sleep, and got the gripes, very severe, and ulcers in your mouth, and you couldn't think clearly."

Nick looked at him, startled.

"How'd you know that?"

"Oh, it's part of my job, I guess. I'm a doctor, occupational medicine. But, tell me, why didn't you quit?

Nick snorted, with a sardonic laugh.

"I guess I just didn't know," he said. "I had no idea what was going on. Nobody told me I was working with a dangerous substance. The guy I bought the business from wasn't there long enough to get into trouble, and I suppose the drugstore chain I was working for was concerned only with getting their thermometers and not with how they were made. And I had far too small an operation for the labor inspectors to bother."

"So what happened?" said David.

"Oh, the inevitable, I guess. Things got worse. I got wilder in my head, and had so much tremor I had to devise tricks to hold the equipment. I even began to get tremors in my eyelids and lips. I'd get muscle spasms sometimes while I was trying to do the job. And none of the doctors seemed to know what it was all about. I guess they weren't occupational medical doctors. They gave me medicine for my stomach and pills to calm my nerves, but nobody ever asked me what I did for a living, and nobody ever seemed to know what was the matter with me. Until . . ." and he stopped.

"Until what?" said David.

"Well, I don't exactly know. I remember being in the workshop. I had a batch of thermometers to make, and I'd boiled up a beaker. But then the next thing I remember I was in a strange bed in a strange room. I had no idea where I was or what was going on."

"You mean you lost your memory for a whole day?" asked David, beginning to develop a clinical interest.

There was a dramatic pause. Nick had obviously told the story before.

"No," he said, shaking his head slowly. "For nearly a year!"

"A year!" echoed David, in disbelief.

"Yeah, nearly a year. At least that's what they tell me. I know nothing at all about it. In fact, I don't know very much about the next six months either. I vaguely remember lying in bed with people coming in and sticking needles into me. But there's nothing clear until about 18 months after it happened. From what I know now, it seems that the cleaning woman found me. She used to come in once a week to clean up the floors and tidy up. She was never really exposed to the mercury. She was only in for a few hours a week. But fortunately, the day I collapsed was the day she came in and she found me on the floor unconscious. They say I was in a coma for about 10 months before I began to creep out of it. But I was in this hospital for two more years before I was able to look after myself properly. I've been out now for about two years and everything's more or less OK." He paused. "So now you know."

Nick sat back in his chair looking triumphant, as though he had successfully completed some sort of ordeal.

"I sometimes feel the need to tell someone about it," he said, hesitantly. "I guess you were the one this week!"

"That's quite a story!" said David, by this time totally impressed. "I'm not surprised you want to talk about it. I don't think anyone else has ever had an experience like that. But tell me, have you any idea how much mercury was in the air you were breathing?"

"Well, seemingly the labor inspectors did some tests a week or so after I left. They say the atmospheric mercury level was greater than 1 mg per cubic meter in the general room air. That's 20 times the TLV, if you know what a TLV is."

"Yes, I know what a TLV is," said David, dryly. "And I even know something about mercury poisoning. And let me tell you, you are one lucky guy. You must have lived right somehow, because almost everyone else would be dead."

"Yeah, I know," said Nick, quietly. He grinned and held up his glass, which was shaking violently. "But it left me with a bit of a problem!"

Five Steps to a Healthier Workplace

Step One: Recognize the Need

In the previous chapters we saw that toxic health problems can exist even in well-organized plants, and that employees, even with the best intentions of management, can suffer from chemically induced illness. We looked at the nature of the chemistry involved, how chemicals can get into the body, and what can happen to them once they get in. We examined also the basic nature of chemical toxicity and how to measure it. When a problem exists, clearly the next step must be to control or, better yet, eliminate it. Unfortunately, the latter is seldom feasible. Generally, the best we can do is establish a degree of control.

Bearing all that in mind, and recognizing that a healthy workplace is a productive workplace, how can one go about developing a still healthier workplace in a logical, sequential, cost-effective, comprehensive, and yet simple manner? When all is taken into account, and all the advice and guidelines are crystalized, five relatively simple steps can be taken. These are as follows:

Step One: Recognize the Need

Step Two: Define the Problem(s)

Step Three: Evaluate the Risk
Step Four: Establish Permissible Limits of Exposure
Step Five: Control the Exposure by
  • Revising Administrative Procedures
  • Applying Engineering Principles
  • Prescribing Personal Protective Equipment

Each one of these steps will be considered in detail in subsequent chapters. It is not the intent here, however, to delve into the complexities and control of esoteric chemicals. What is intended is to show how an ordinary person, who is not a specialist in occupational health or occupational hygiene, can examine the conditions in his or her plant, determine if there are indeed any health problems, and either deal with them or recognize that the assistance of a specialist is required. But in the words of the song, no one is promising a rose garden. Even if you follow every precept and dictum presented here, you will still have problems. That's the nature of work. But you will also have a healthier workplace and a more satisfied workforce, two elements of management which, when other things are equal, almost inevitably lead to greater productivity.

Before discussing Step One, however, two considerations must be taken into account. The first is the concept of a healthier workplace. What we are trying to do here is to show how one can go about improving the health potential of the workplace by eliminating or reducing exposure to toxic materials. We are not concerned here with the concept of safety in the traditional sense nor with exposure to physical hazards, but with something less tangible, although equally important, namely, the prevention and control of occupational illness. The second consideration is an extension of the first, namely, that in working towards a healthier workplace, one has to accept the fact that elimination of toxic exposure is not always feasible, nor even possible. As we shall see when we examine Step Four, some level of tolerable exposure to hazardous chemicals sometimes has to be accepted, provided it is indeed tolerable and relatively harmless for a normal working life.

HOW DO YOU RECOGNIZE THE NEED?

So how do you know you've got a problem? Maybe it's easy. Maybe you know someone on the shop floor who has skin troubles from solvents or, worse still, maybe someone with ailments like Jake Whitman's works for you and you have problems you don't even want to think about. Perhaps it's obvious that something is wrong because everyone is getting sick, but you do not know what the cause is.

The first thing to do is to recognize that there *might* be a problem in your particular plant, even when conditions are apparently good. Remember Mrs. Madison and the women in the secretarial pool. It's very easy, particularly when conditions are apparently good, to ignore the fact that things might not be as they seem. Step One, then, is to recognize the need.

There's an old adage that says if it ain't broke don't fix it, and that certainly is true. But you can't even make the decision to fix it unless you know whether it's broke in the first place. That may seem self-evident, but many people, managers and others, may be so concerned with other legitimate matters that they don't immediately recognize the need, even when it is there. If you want to recognize whether there is a need, you have to start by asking yourself, objectively and with an honest answer, is *my* workplace unhealthy? Do *I* have cause for concern? Do *I* really know for sure? Should I be suspicious of some possible problem areas in the plant, or am I justified in thinking that because my plant has passed the government inspections and nobody has complained too much then I can't have any health problems? These questions are not always easy to answer. You may be aware of problems but haven't gotten around to fixing them yet, or you may know you have problems but you don't know what's causing them, or indeed there may be problems around that you're not even aware of. So if you want to know what to do, how do you go about it?

The answer is really quite simple: Look, Ask, and Listen. Look with new eyes, ask with fresh questions, and listen with willing ears. Perhaps that appears obvious, but sometimes the apparently obvious needs to be reiterated from a different point of view for one to see it more clearly. Let's look at the workplace with different eyes and see what might have been missed before.

If you are already aware of the existence and general nature of one or more health problems in your plant, then to a greater or lesser extent you have already taken Step One. But, if you are not too sure what your problems are, or even if you have any, there are several ways to find out. Even if you think you don't have any problems, it's still worthwhile to take a look, because you might be able to find and fix something before it develops into a more serious situation.

The simplest way to begin the process is to start looking for the presence of patterns or clusters of health complaints or even clusters of specific sickness absenteeism. Statistically, in a random world, clusters of apparently related items can occur strictly by chance. Nevertheless, when you find a cluster or patterns of health complaints, it's worth looking into to see whether there is any apparent cause-and-effect relationship.

If you can find one, you have a starting point to determine your future action, which will be detailed later. However, even if you can find some clusters, but can't find a simple cause-and-effect relationship, the presence of the clusters will suggest that something is wrong somewhere and it may need to be fixed.

Thus the process of Step One becomes a series of sub-steps, as follows:

- Define patterns or clusters by
  - Examining records, reports, and complaints
  - Talking with knowledgeable persons
- Identify potentially hazardous chemicals
- Consolidate the relevant information

EXAMINE RECORDS, REPORTS, AND COMPLAINTS

ABSENTEEISM RECORDS

All plants maintain records of absenteeism, if only for reasons of keeping track of pay. These records, however, can provide a valuable source of information about the health of the workplace. If you trace them back over a few months, you will soon establish whether the absenteeism is distributed fairly evenly throughout the whole workforce or whether most of it seems to be coming from one or more specific groups or work areas. When you find that there is more sickness absenteeism among employees in a certain operation or area, your suspicions should be aroused that some factor in that operation or area could be giving rise to the excess absenteeism.

What you should look for in these records, then, is the presence of some trend or pattern. You might find, for example, that a disproportionate number of employees from a given operation or area within the plant has been receiving medical treatment over the past few months or even years. You might not know the reason for the treatment, but the mere fact that a disproportionate number of workers has been treated should arouse your suspicions. If you looked further, you might find that many of the employees in that area are being treated for some specific complaints such as sore throats and cough. These complaints perhaps might not be sufficiently serious to cause significant lost-time injuries, but they might be serious enough to interfere with best performance or even well being. And if you found that work in that area involved the use of some particular solvent or some breathable chemical, and that the complaints dated from the time these were first used or the time when the workers concerned moved into that area, then you could be suspicious of some cause-and-effect relationship. Similarly, if you found a lot of employees in a particular job were getting skin conditions, then you might suspect that some other irritant chemical might be the culprit. If many employees seemed to be getting recurrent stomach aches or digestive upsets, you might suspect a lead or other heavy metal problem, and so on. Any situation with an unusual cluster of complaints should be looked at. Sometimes you can find a link, sometimes you can't. When you do find a link, then you've got something to go on. But, even if you can't establish any cause and effect, you will still have isolated a problem that requires further attention.

HEALTH RECORDS

What you want to do at this stage is get access to health records to confirm or deny some of this cause-and-effect relationship. This, of course, and rightly so, is not permissible. Health records, even those developed and maintained by company health personnel, are the personal and private property of the employee, and even if you are fortunate enough to have an occupational health service in the plant staffed by trained occupational health personnel, you cannot by law have access to personal health records without the express permission of the employee concerned. That doesn't mean you can't get useful information, however. The doctor, nurse, or other health workers can legally extract nonconfidential information from the records and send the results to you. In fact, this should be the normal procedure, if you have an occupational health service. Indeed, there are few persons in a plant who are in a better position to establish the existence of health problems in that plant than an occupational health nurse. A good occupational health nurse does not, of course, merely spend his or her time in the first-aid room or the health department applying slings and bandages and dispensing aspirin and kind words. She (or he) organizes health care in the plant, establishes record systems, develops habilitation and rehabilitation programs, liaises with exterior medical facilities and personnel, and so on, but also spends time on the shop floor, looking for, talking about, and listening to health issues in a much more productive way than can ever be achieved by even the most conscientious of line supervisors or staff managers.

However, even if you don't have a health service, it is still possible to get sickness and injury information from the local doctors, clinics, and hospitals where your employees are treated. Again, this cannot be personal confidential information, but with suitable liaison, you can often get enough to establish trends, even by simply asking the local doctors and health facilities personnel if they have become aware of any trends in their findings with your employees. In fact, with the concurrence of the union, if there is one, it is sometimes possible to set up a health record system in the human resources office that does not include confidential information but does provide trend information.

Doctors, of course, are oriented towards diagnosis and treatment. They are interested in treating the sick or in curing a particular disease or injury. Unfortunately, they have little or no knowledge of what goes on behind the plant doors, and even when a patient may suspect some factor at work might have caused his complaint, the explanation may have little meaning to the doctor. As a result, the doctor may have little or no understanding of what caused the problem in the first place, and little interest in collecting data linking the source with the condition. To alleviate this problem, some plant managers or human resource managers have found success in inviting local doctors into the plant to see what goes on so they can have a better understanding of factors that may be causing their patients' conditions. Some, indeed, have gone so far as to combine a meeting of the local medical association with a cocktail party or other welcome activity at the plant.

Employee health records, of course, are not the only source of valuable health knowledge in the plant. Workers' compensation and insurance records, although limited in number, can be even more useful if they are available.

WORKERS' COMPENSATION AND INSURANCE RECORDS

In those jurisdictions where a formal workers' compensation system is in effect, the compensation health records provide the most definitive information available because the medical, social, and work history of each employee who applies for workers' compensation is investigated in comprehensive detail and is made available in large part to management. Currently, relatively few illnesses, as compared to injuries, are considered for workers' compensation claims. However, if a cluster can be found with the same type of complaint, that information can be invaluable. Similar information can be obtained from private insurance companies who provide employee coverage in place of workers' compensation.

LABOR INSPECTORS' REPORTS

Labor inspectors' reports often tend to be dismissed as documents to be tolerated and summarily dealt with or even ignored if possible, rather than as sources of valuable health information. However, some very useful information on health trends can be derived from them over and above the demands that they may present. An inspector looks at plant operations with the eyes of an informed stranger, and because of that, he or she can see things that someone who is familiar with the operation might overlook just because of that familiarity.

Thus, an inspector's report can reveal situations that are quite obvious when they are pointed out, but might never have been noticed. Furthermore, some plants, especially the larger and/or more intrinsically hazardous, receive a lot of inspections; others little or none. When there are numerous visits, the sequence of information in these reports, if examined with a critical eye, may demonstrate the existence of hazardous or potentially hazardous areas or operations that even an inspector might have missed in a single report. Thus, in trying to establish the presence or absence of health trends, it is wise to consider in your evaluation the information the labor inspectors have recorded about the operation.

COMPLAINTS AND GRIEVANCES

Complaints and grievances are still another source of information. Much can be gained by studying the overall nature of complaints and grievances to determine whether they too present any pattern or trend. Informal complaints may arrive anonymously in a complaint box or may be presented directly by the employee to someone in authority. Regardless of their origin, it is useful, if the practice is not already in effect, to keep a register of these complaints so that patterns and trends can be observed. When complaints become grievances then, of course, the matter becomes formal and a full investigation may reveal some significant underlying cause that perhaps was not previously known and might otherwise remain a smoldering health problem.

HEALTH AND SAFETY COMMITTEES

Most jurisdictions today require that companies employing more than a certain number of workers maintain a committee of management and labor representatives to conduct some form of regular health and safety inspection of the workplace and submit regular reports to management. The effectiveness of these committees varies considerably, depending on such things as the enthusiasm of management support, the dedication of the worker representatives, and the participation of the general workforce. The reports, however, particularly those from an effective committee, can provide information that is not available from any other source, and can point out conditions and situations that might not otherwise be evident. They, too, should be studied to reveal whatever trends they might expose.

TALK WITH KNOWLEDGEABLE PERSONS

One of the simplest ways to find out what's going on with the health of your employees is merely to ask questions. Speak to anyone you can find who might have something useful to say. If you're in production, speak to human resources and health and safety personnel. If you're in personnel or health and safety, speak to production people. Speak to employees on the shop floor, to health and safety union representatives, if any; speak to safety coordinators, health services personnel, supervisors, anyone at all who is involved with the operation. Ask them if *they* have any suspicions about health and safety on the shop floor. And listen to what they say. Don't dismiss the answers because you feel the respondent might be uninformed or prejudiced. And don't stand on authority. You're not inspecting the troops. You're trying to get information to make things better for everyone, and ultimately to improve the bottom line. At the same time, you don't have to accept everything that is said, but remember that the worker on the shop floor is closer to the action and may know more about some aspects of working conditions than anyone else. And remember, too, that even a manager is entitled to some humanitarian feelings.

IDENTIFY HAZARDOUS MATERIALS

Let's assume that the records you examined and the discussions you held indicate that a health problem exists somewhere in the plant and it seems to be chemical in origin. One of the first things you can do to find out what's causing the problem, and thereby recognize a need for further action, is to identify all the potentially hazardous materials in the working environment. This may sound like a huge task, and sometimes it is. But, as a matter of fact, it should have been done already. In many jurisdictions, it is mandatory to file an inventory of all hazardous materials in the workplace and to make their potential hazard known to the workers. But, if it hasn't been done, or if what has been done is inadequate, then this is a good time to start. To do it systematically, however, you must consider the whole process, starting with the primary raw materials, continuing with any materials that are added in the course of the process, as well as chemicals that are used in some support function such as solvents or cleaners, and end products and by-products, and those chemicals that have been put in storage for occasional use or whose use may have even been forgotten. Thus, each chemical used in a given operation, or stored for future use, or held as inventory for distribution has to be considered in terms of its potential health hazard.

One of the best ways to determine that potential is to examine the Material Safety Data Sheets (MSDS), which by law in most jurisdictions must be provided by the manufacturer for any hazardous material used in the workplace and made available to the employees by plant management. The MSDS provides all the information necessary to determine the hazard potential and the requirements for control of any hazardous substance. Still other information can be found in the various guidelines published by the American Industrial Health Association, or by the U.S. National Safety Council, or by NIOSH (the U.S. National Institute for Occupational Safety and Health) or various other national, state, or provincial government bodies. (See Appendix II for information on sources.) Once you have an inventory, of course, you can start relating potential health problems to possible sources in particular areas. Also, in the process of preparing this report, you will begin to build an invaluable library that will stand you in good stead for further consideration of the problems at hand.

CONSOLIDATE THE RELEVANT INFORMATION

The previous sections have demonstrated that relevant information on which to base a recognition of need can be derived from widely different, but relatively accessible, sources. Information from absenteeism records, health records, and workers' compensation or insurance records can help establish statistical trends. The presence of clusters, or trends, can indicate the existence of possible health problems. Information from complaints and grievances and discussions with employees can suggest where problems might lie. The existence of potentially hazardous chemicals might suggest where the origin of these problems could be. Before too long, by studying these various records and talking with the persons involved, you'll begin to establish a picture that will show you that you have nothing to worry about, or you do have something to worry about even though you don't yet know what it is, or you know what it is, but you don't know how to fix it.

If there is something to worry about, Step Two in the next chapter will explain how you can go about defining it.

CHAPTER FOURTEEN

Step Two: Define the Problem

BACKGROUND

Now that we have seen how to recognize the need for further action, we must consider where to go from here. The ultimate objective, of course, is to eliminate or at least control any health problems that could arise from the presence of toxic materials in the environment. Before you can begin to exert any control, however, you must define the specific problems. You can't just vaguely accept the fact that there's something wrong. You can't even consider what to do until you know what it is you are dealing with. In other words, you must establish clearly in your own eyes, or perhaps in the eyes of higher management, what the nature of the problem or problems might be.

Although this may be a simple process in principle, it is not necessarily so in practice. Nor is it necessarily only a function of management. Labor activities, government regulations, and social pressures demand that employees as well as employers become involved in the health and safety of the workplace. Labor unions require assurance of health at work and are not necessarily satisfied that management alone is capable of doing the task. Legislation may demand regular formal assessments to determine to what extent chemical and other health hazards may exist. However, this is not to say that you as a representative of management should not be doing the job. Indeed you should, but ultimately legislation and social pressures will ensure that labor, whether organized or not, will inevitably be involved as well. However, if you as a manager are clear in your own mind about the state of health conditions in your workplace, you can more easily deal with any external pressures that may be brought to bear on you. Keep in mind that a curious paradox sometimes exists in industry today. Because of training courses presented by unions and other bodies, the employee on the shop floor may in some instances be more knowledgeable about health hazards in the workplace than management and may not be averse to displaying that knowledge.

To define the problem or problems, then you need to know exactly what's going on with health conditions in the workplace. In other words, you need to make a formal assessment of the health status of the workplace. You've done the preliminary work as described in Chapter 13. You've examined the records and the statistics; you've established the prevalence and causes of absenteeism; you've inventoried the workplace chemicals; you've talked with everybody who might have something helpful to say. Now is the time to bring all of this information together and apply it to the shop floor. One of the best ways to do this is to conduct what is called a walk-through survey.

WALK-THROUGH SURVEY

A walk-through survey is not just a casual amble through the plant to see that everyone is performing properly or wearing the appropriate piece of safety equipment. Instead, it is a systematic examination of all relevant plant activities to determine, define, and record the presence and extent of current health hazards and to estimate the likelihood of potential health hazards arising from the presence of toxic materials. A walk-through survey can, of course, be concerned with other matters besides those pertaining to toxic exposures. Other hazards, physical, ergonomic, and biological, are also of concern, but at this time we are dealing only with those that derive from toxic exposure.

WHO SHOULD CONDUCT THE SURVEY?

Large corporations may have specialized personnel such as occupational hygienists, health-trained safety officers, and occupational health nurses, who have received specific training in conducting surveys of this nature, and if they are available obviously they should do it. Most organizations, however, don't have that luxury. For the smaller corporation or plant without specialized personnel, and particularly for a preliminary screening, the responsibility must rest with the owner, the plant manager, the production manager, the human resource manager, or whoever happens to be the most reasonably accountable. There is no reason why a preliminary walk-through survey cannot be satisfactorily conducted by any responsible person who is knowledgeable and has the ear of someone who can get things done. It is the purpose of this book to ensure that the person responsible is in fact knowledgeable.

PREPARATORY CONSIDERATIONS

Before beginning the actual detailed examination that is the thrust of the walk-through survey, it is necessary to prepare yourself for what you are about to do. You can't just walk in and start looking at various activities. If you are not already familiar with every aspect of plant activity, you should familiarize yourself with the nature of the operation, including the organizational and demographic background, the processes, the tasks, the flow of activities, the equipment, and so on. Before you start any inspection you should ask yourself some specific questions such as those listed in Table 14-1.

If you've done your homework and are completely familiar with the plant and its activities, then these preliminary questions can be bypassed, although answering these questions is sometimes useful just to remind yourself of the background. If you are not thoroughly familiar with the actual operations of the plant, then the next step should be to take an introductory inspection of the operations you are going to be concerned with. This introductory inspection is not the detailed examination of the walk-through survey, but is in the nature of an overview. This type of inspection is particularly valuable for someone who is not completely familiar with the specifics of the operation such as a human resources manager who has been pushing more paper than he or she would like or even an owner or production manager who doesn't get on to the floor too often. This introductory inspection then takes the form of a continuous observational walk

Table 14-1
Preliminary Questions Before a Walk-Through Survey

Demographic and operational information
- What specific operation(s) is/are to be examined?
- Who is responsible for the operation(s) to be examined?
- Have you received their authority, if necessary?
- Who are significant contact persons for further information?
- Who are the worker health representatives (if any)?
- How many shop floor employees are involved?
  Male? Female?
- What trade unions (if any) are involved?
- What is the incidence of sickness and other absenteeism in the area of interest?
- Are there observable patterns of absenteeism, injury, and need for medical care in that area?
- What is the incidence and nature of relevant cases reported to workers' compensation or insurance agencies?
- What does management perceive as the main health problem?
- What do the workers perceive as the main health problem?

Work organization pertinent to health
- What are the company operational policies and procedures with respect to health in the workplace?
- What are the laws and regulations pertinent to the operation(s) of concern?
- What, if any, is the system of health protection and surveillance, with respect in particular to:
     pre-employment and routine physical examination?
     special medical and biological testing?
- Are the supervisors familiar with the needs of good health and hygiene, and do they implement good practices?
- What, if any, is the system of formal job rotation?
  Does it reduce exposure to toxic materials?
- Are health complaints adequately investigated and are corrective procedures applied where necessary?
- Is formal and repeated training for maintenance of health in the workplace provided?
- Are instructions for healthy work practices put in writing?

Source: Derived in part from Fraser, T. M. The Worker at Work. London, New York Philadelphia: Taylor & Francis, 1989, pp. 387, 420.

through the plant or the area of concern from the point of intake of raw materials to the output of finished product. If you don't know the technical aspects of the process, you should make sure you are accompanied by an expert who does, although that expert doesn't need to be a member of management. At this stage, you should not stop to examine each process in detail. All you are trying to do is to get an overall picture of what's going on. During this walk, of course, you will confirm the previously determined relationships and process flows and take mental note for future reference of points of potential hazard and areas or processes for special future consideration. On completion of the preliminary walk, you should have a clear overview of the conditions and processes involved and should be able to define more clearly the various activities, processes, or areas where excessive exposure might be found so that you can return to them for the next stage of detailed examination. You may also find it useful to have still further discussion with others in the plant or collect additional information from reference sources to clarify points of uncertainty.

DETAILED EXAMINATION

The detailed examination, as the name implies, is a close inspection of the various processes that your preliminary examination indicated to be of concern. The usefulness of this detailed examination hinges on your ability to use your basic senses, in particular, by looking, smelling, listening, and perhaps, where appropriate, even touching, and to confirm your suspicions by talking with the persons involved, including both management and labor, and conducting such tests as may be necessary or helpful.

In seeking out health problems, what is it you should be looking for? The glib answer is anything at all that would seem to be a potential health hazard. To make that answer a little more specific, however, it is useful to consider a formal checklist or questionnaire, such as that presented in Table 14-2.

You should note that some of the questions refer to material on evaluation and control, which will be discussed in Steps Three, Four and Five. By the time you come to use the checklist, however, you will be familiar with the content of these various steps and will be able to apply the knowledge in your survey.

A checklist, particularly a comprehensive one, can be used in various ways. You can use it as you walk through the plant, to check off each item as you go and you can repeat the process as you approach a new area or activity. However, this approach is cumbersome, tedious, and impractica-

Table 14-2
Questionnaire for Use During a Walk-Through Survey

Materials

What is (are) the primary materials?

(e.g., chemicals, part finished components)

What are the end products?

(e.g., chemicals, manufactured articles, hardware)

What other materials are added or used during processing?

(e.g., special chemicals, solvents, degreasers)

What are the by-products?

(e.g., other chemicals)

What are the effluents and residues?

(e.g., gases, mists, fumes, dusts)

Processes and Equipment

What are the processes?

(i.e., describe what happens. Prepare rough flow chart sketch.)

What equipment is used in processing?

(e.g., reactors, presses, grinders, furnaces, and tools)

Hazards

Are chemicals used in the operation that have the potential for:

• toxicity?

• flammability?

• explosivity?

• inadvertent reactivity?

Does that potential (if any) exist in:

• primary materials?

• added materials?

• end-products?

• by-products?

• effluents and/or residues?

Does any potential hazard occur in connection with:

• storage of raw materials?

• transport of raw materials?

• adding or mixing of raw materials?

• processing materials?

• transport of by-products and end-products?

• storage and distribution of by-products and end-products?

• storage and/or elimination of effluents and residues?

(continued on next page)

Table 14-2 (continued)
Questionnaire for Use During a Walk-Through Survey

What is (are) the probabilities of exposure?
(Specify likelihood)
Is the exposure local at the site or general throughout the plant?
(Specify)
How many employees are exposed?
(Job category? Name?)

Evaluation and Control

What sampling, monitoring, and /or measurement procedures are in
effect at the site(s) or in the area? (Specify)
Is there need for further sampling, monitoring, or measurement?
What control procedures are in effect at the site(s)?
- administrative
- engineering
- personal protective
What, if any, ventilation system is used?
- general
- local
Is the ventilation system, if any, adequate with respect to design and
use of:
hoods?
fans?
air cleaners?
recirculation systems?
replacement air?
Is there need for new or additional control procedures, including
ventilation? (Specify)
Are any processes or operations partially or fully enclosed? (Specify)
Is a need for new or additional enclosure apparent? (Specify)

Housekeeping and Hygiene

Are hazardous materials stored according to best practice?
Are containers covered when not in use?
Is there vacuuming or other cleaning of dust?
Is the floor area clean and free from clutter?
Are eating facilities provided away from the shop floor?
Are adequate washing facilities provided?

(continued on next page)

Table 14-2 (continued)
Questionnaire for Use During a Walk-Through Survey

Are eating and smoking prohibited on the job?

Are clean work clothes provided? If not, are work clothes kept free from impregnation or soaking with hazardous materials?

Is it desirable, and feasible, to change an existing hazardous process for one less hazardous?

Is it desirable, and feasible, to introduce wet methods for dust control?

Is it desirable, and feasible, to improve the general level of house-keeping and hygiene? If so, how?

Source: Derived in part and further developed from Fraser, T. M., The Worker at Work, London, New York, Philadelphia, Taylor & Francis, 1989, pp. 422, 424.

ble. It is much more realistic and cost-effective to thoroughly familiarize yourself with the principles and content of the checklist before you begin the walk-through and then conduct the survey with the content and principles of the checklist in mind. On completion of the walk-through, you can refer back to the checklist to see what you might have missed, and, if necessary, go back and have another look.

In the course of the survey, in addition to what you actually see, you should pay attention to the unexpected. A container, for example, properly processed, properly sited, and in all respects satisfactory, could unexpectedly rupture and spill its contents. Or, a toxic effluent extracted to a roof vent through an otherwise effective ventilation system (as we have seen with the tale of Mrs. Madison) could be pulled back into circulation through an improperly located air intake. One of the keys to the survey, in fact, is alert awareness.

RECORDS

A survey of this kind, however, no matter how thorough, is of no value to anyone but yourself if the findings are not recorded. As you go around the plant you should carry and use a notebook, scratch pad, or portable tape recorder to record your comments as you find them. This record need not be anything more than field notes that include the date, time, and location of the process or area you are specifically examining, with brief comments on what you consider important in the light of your pre-established knowledge and the demands of the checklist. It is highly desirable, however, to include sketches, Polaroid or other photographs, and, where the dynamics of the process are significant, a video tape record. Be sure, of course, that you correlate your sketches and photos or video records with your written notes.

After completing your survey, re-examining the checklist, and perhaps reconsidering certain areas, these notes and visual records can be organized, evaluated, and edited as a formal report complete with recommendations for submission if necessary to higher management. Even if you don't have to submit this report to higher management, it is wise to go through the same process and keep the resulting report as a guide to future action or even as a document for use should future legal complications arise.

What can you get from your survey? Apart from a thorough knowledge of the overall health aspects of your plant or whatever area concerns you, you are now in a position to define with reasonable probability the source or sources of whatever health problems you might have or, at least, to determine what further action you should take to transform that reasonable probability to reasonable certainty.

We have looked then at Step One, Recognize the Need, and Step Two, Define the Problem. We now have to determine to what extent any problem or problems we have might constitute a risk. This brings us to Step Three, Evaluate The Risk, which we shall consider in the next chapter.

Step Three: Evaluate the Risk

HAZARD, RISK, AND DANGER

As we have noted, there are tens of thousands of chemicals in the workplace. Many of them are toxic. But when they are used safely, or in small enough quantities, they don't present any danger. Thus, the mere existence of a toxic chemical doesn't mean that the persons exposed to it are in danger.

Three terms must be considered before evaluating a possible hazard: hazard, risk, and danger. The word *hazard* refers to the *existence* of a potentially harmful condition. For example, if you're working with or near lead or carbon monoxide, you are in a potentially harmful situation, but are not necessarily in danger. The term *risk* describes the *probability* of harm, or the extent to which you are likely to be harmed. If the carbon tetrachloride you're working with is in a properly ventilated and closed tank and you don't have any contact with it unless while wearing gloves and/or a respirator, the risk is minimal. It is possible to measure the risk in terms of, for example, the likelihood of being harmed during exposure to a given concentration of a chemical for a normal working week. This is the principle known as *threshold limit value*, which will be discussed later in this chapter.

Danger defines the state in which the risk reaches an unacceptable level. The term unacceptable, of course, is subjective. Sometimes it is easy to define, as in imminent danger of death, sometimes difficult, and indeed defining danger may become something of a judgment call in the light of what the public perception may be. For example, the chemicals known as isocyanates (MDI and TDI), which are used in making plastic foam, can give rise in susceptible persons to a sudden and disabling asthma that can be permanent, even with minute quantities. Exposure to these chemicals thus becomes totally unacceptable. Tolerance is essentially zero. The manufacturer using MDI and TDI must take exceptional protective measures to minimize any human exposure. On the other hand, exposure to a nuisance dust such as carbon black, which is dirty, uncomfortable, and even thick to breathe, does no apparent harm except to blacken your lungs and skin. Although it has no demonstrated health hazards, it also must be considered unacceptable because its presence can be esthetically intolerable.

FACTORS TO CONSIDER IN EVALUATION

CHARACTERISTICS OF PERSONS EXPOSED

Not the least of the factors for consideration in evaluating a possible risk pertains to the characteristics of those persons exposed. A hazard does not become an actuality until there is a risk of human exposure. Consequently, in evaluating the risk, you must take into account those who are or might be exposed, not only in terms of numbers but also in relation to the realities of individual variation and predisposition. These considerations can be difficult in a political environment that demands equality. However, it should be recognized that not all persons are equal either in their predisposition towards an adverse response to a toxic assault or in the severity of their response to that assault. Differences in response can occur, for example, by reason of age, sex, and physical fitness. A classic example exists in the manufacture and processing of female endocrine hormones in which a woman, and particularly a pregnant woman, may be more at risk than a man under the same circumstances. Less dramatic, but

even more significant, is the fact that tobacco smokers are more susceptible to damage from other chemicals by inhalation, particularly inhalation
of particulates such as asbestos. Heavy alcohol drinkers are more susceptible to liver and brain poisons such as those found in various solvents,
and drug users in general tend to respond more adversely than their drug-
free fellows. Taking these predispositions into account may be difficult in
a normal working employee-management relationship, and sometimes it
would not be practical as in the case for example, of insisting that only
non-smokers work in a particular environment. But at least the situation
should be kept in mind.

It is also necessary to bear in mind the possible outcome of an exposure. When the worst outcome is likely to be a minor disability, such as an
irritating cough or an annoying skin eruption, then the risk is minimal, but
it should still be reduced as close to zero as is feasible. On the other hand,
when the worst outcome is a major disability, such as debilitating
bronchial asthma, liver and kidney disease, destructive blood disease,
brain or nerve damage, cancer, or untimely death, then hazard control is
absolutely necessary. It becomes essential, then, that you as a manager
know what the possible outcomes are, how to establish permissible limits
of exposure, and ultimately how to control the real and potential hazard.

NATURE OF RISK

Risk, then, can refer to:

- the probability of occurrence of a toxic exposure
- the probability that a person may suffer an occupational illness as a
  result of that exposure, or
- the chance a worker is willing to take to enjoy the benefits of his/her
  employment.

In *Product Safety Management and Engineering,** Willie Hammer
points out that to be perfectly safe a "product," or in this case, a "chemical agent," would have to permit no possibility of causing injury or damage under any circumstances—a situation that is nigh to being impossible.
A person therefore must expect some minimal element of risk. How much

---

*Hammer, Willie. *Product Safety Managment and Engineering.* Englewood Cliffs, New Jersey: Pren-
tice-Hall Inc, 1980.

risk depends on the cost/benefits to the individual and to society. The actual point at which a risk becomes unacceptable remains to be defined.

A risk can be voluntary or involuntary. A voluntary risk is one in which the individual participates in an activity by freely accepting the risk based on criteria from that person's own values and experiences. An example of this is hang-gliding. An involuntary risk involves participation in an activity that is not, or only partially, under the control of the participant. At one time, a worker was expected to accept the inherent risks of a job if he/she undertook to do that job. That concept—the assumption of risk doctrine, as it was called, no longer applies. In today's society, the company, the corporation, or just the boss if the plant is small enough, is normally held responsible in law for providing a working environment that is safe and healthy. Indeed, as Hammer also points out, the worst course of action for a company is to disregard a discovered or reported hazard. A company could be held legally responsible if a legal suit were to be brought against them in which someone was severely injured or killed because a hazard had been ignored. If a company is aware of a problem, especially a recurring problem, failure to take corrective action can lead to the award of punitive damages that are far greater than the cost of taking corrective action.

RELATIVE RISK INDEX

How can one analyze the extent of the risk? How can one determine to what extent it is acceptable or unacceptable? It is possible, as noted earlier, to use complex mathematical probability analyses to achieve this end. On the other hand, one can use a simpler, more practical approach, such as the following, to develop what might be termed a Relative Risk Index.

This approach is based on concepts developed by the U.S. Air Force to define the extent of possible safety hazards. The approach applies qualitative ratings to a series of questions about the chemical environment.

The rating table presented in Table 15-1 and the questionnaire in Table 15-2 can be used for this purpose.

The table is made up of three parts. The first part defines four significant categories of a potential hazard, ranging from the lowest (negligible) to the highest (catastrophic). The second part defines the likelihood of hazardous exposure in qualitative terms, that is, in six levels from frequent to impossible. A numerical rating from 6 (frequent) to 1 (impossible) is applied to each term. The third part combines the significance rating with the likelihood rating in a simple matrix.

Table 15-1
Toxic Hazard Rating Table

Significance	Description
I	Negligible: No injury or occupational illness
II	Marginal: May cause minor injury or illness
III	Critical: May cause severe injury or illness
IV	Catastrophic: May cause death

Likelihood of Occurrence	Rating
Frequent	6
Reasonable	5
Occasional	4
Remote	3
Extremely improbable	2
Impossible	1

Rating Matrix

Likelihood	Significance	I	II	II	IV
1		1	2	3	4
2		2	4	6	8
3		3	6	9	12
4		4	8	12	16
5		5	10	15	20
6		6	12	18	24

Any combination with a matrix rating of 9 or more is considered to offer an unacceptable risk.

The questionnaire in Table 15-2 can be used with the information derived from the walk-through survey you conducted in Step Two in the previous chapter. You can now determine potential hazard levels in the shop by applying the rating levels from Table 15-1 to the answers to the questionnaire in Table 15-2. Here's how it works:

- Using the information you collected with your walk-through survey, provide specific answers to each question in Table 15-2.
- Using Table 15-1, apply a significance rating and likelihood rating to each of your answers.
- Calculate the combined rating by multiplying the significance rating by the likelihood rating.

Table 15-2
Rating Questionnaire

Topic	Significance	Likelihood	Combined rating
1. Is there a potential hazard with: a) primary materials? Specify.			
b) end products? Specify.			
c) added materials? Specify.			
d) by-products? Specify.			
e) effluents & residues Specify.			
2. Are processes likely to fail with possible toxic exposure? Specify which.			
3. Is equipment likely to fail with possible toxic exposure? Specify which.			

(continued on next page)

Table 15-2 (continued)
Rating Questionnaire

Topic	Significance	Likelihood	Combined rating
4. Is ventilation prone to fail with possible toxic exposure? Specify where.			
5. Are storage practices likely to generate toxic exposure? Specify where.			
6. Are distribution practices likely to generate toxic exposure? Specify where.			
7. Are housekeeping and hygiene practices likely to generate toxic exposure? Specify.			

The resulting combined rating, or Relative Risk Index, is considered to be unacceptable if the level is 9 or above. However, these ratings are not etched in stone. They are merely a guideline. Under certain circumstances, a risk may be unacceptable even if it does not meet the above criteria.

We have seen, then, how to recognize when a problem exists, how to define the problem, and how to evaluate the associated risk or risks. The next step is to define what level of exposure can be considered permissible in an industrial environment.

CHAPTER SIXTEEN

Step Four: Establish Permissible Levels of Exposure

EXPOSURE LEVELS

As we have seen, toxic materials are commonly found as aerosols, that is, floating minute particles within the breathing air, either as gases and vapors or liquids and solids. As we have seen also, gases and vapors are measured in volume units, namely, parts per million (ppm), while liquids and solids are measured in weight units, namely, milligrams per cubic meter.

To some people, the concept of an acceptable limit of exposure to a toxic material is anathema. Nevertheless, in recognition of practical realities, various authorities, federal and state in the U.S., and federal and provincial in Canada, and in jurisdictions elsewhere, developed guidelines for the establishment of acceptable limits of exposure. In particular, OSHA, the Occupational Safety and Health Administration in the U.S., has promulgated standards that must be adhered to, while various other government bodies have done likewise. All the guidelines, however, including those throughout most of the world, derive from the original work of the American Conference of Government Industrial Hygienists (ACGIH) who developed a concept they call *Threshold Limit Values* (TLV). A Threshold Limit Value, which is a proprietary term, refers to the airborne concentration of a substance and represents conditions under which it is believed that nearly all workers may be repeatedly exposed to that substance day after day without adverse effect. The booklet, *Threshold Limit Values for Chemical Substances and Physical Agents in the Working Environment,* published under copyright by the ACGIH (See Appendix II for the source) lists TLV's for more than 1,200 common industrial chemicals. These threshold limits are based on the best available information from industrial experience, from experimental human and animal studies, and, when possible, from a combination of all three. The basis on which the values are established may differ from substance to substance. Protection against impairment of health may be a guiding factor for some; whereas, reasonable freedom from irritation, narcosis (stupor), nuisance, or other forms of stress may be the basis for others.

The amount and nature of the information available for establishing a TLV also varies from substance to substance. Consequently, the precision of the estimated TLV is also subject to variation. The ACGIH specifically points out that TLV's should not be used as a relative index of hazard or toxicity. However, they commonly do express some element of relative hazard. In particular, however, they should not be used to distinguish fine lines between safe and dangerous concentrations. The best practice, of course, is to maintain concentrations of atmospheric contaminants as low as practical at all times.

The TLV is not a single entity. The ACGIH defines three categories of TLV, namely, the *time-weighted average (TLV-TWA),* the *short-term exposure limit (TLV-STEL)*, and the *ceiling.* Because these are proprietary terms, the U.S. Occupational Safety and Health Administration (OSHA) has adopted different terminology for essentially the same concepts, as follows:

PERMISSIBLE EXPOSURE LIMIT (OSHA PEL)

The OSHA-PEL defines the average concentration permissible to a worker for a normal eight-hour workday and a 40-hour work week to which all workers may be repeatedly exposed, day after day, without adverse effect.

SHORT-TERM EXPOSURE LIMIT (OSHA STEL)

It is permissible to a certain limit for a worker to exceed the TWA for a short period during the work day when it is necessary. The STEL is defined as that concentration to which workers can be exposed continuously for a short period of time above the TWA without suffering from a) irritation, b) chronic (that is, long-term) or irreversible tissue damage, or c) narcosis (that is, reduced alertness leading to unconsciousness) of sufficient degree to increase the likelihood of accidental injury, impair self-rescue, or materially reduce work efficiency. Exposures at the STEL should not be longer than 15 minutes and should not be repeated more than four times per day. In addition, there should be at least 60 minutes between successive exposures at the STEL, and a STEL exposure must be compensated by a comparable low concentration exposure during the remainder of the day.

CEILING

Some substances are considered to be so toxic that an averaging method does not apply. For these specially listed substances, the recommended level, denoted by (C) in the booklet, becomes the "ceiling" level, which is the level that should not be exceeded during any part of the working day.

"SKIN" NOTATION

The ACGIH reference booklet also defines certain chemicals, such as methyl alcohol, which have specific effects on, or can be absorbed through the skin. In these situations the word "skin" is printed after the name.

NUISANCE DUSTS

While some particulates, such as silica, produce well-recognized pathological effects, others, such as carbon dust, are more or less inert. However, there is no dust that does not produce some effect on the lungs when inhaled continually for perhaps years. Excessive concentrations of nuisance dusts in the air may cause reduction in visibility as well as uncomfortable deposits in the eyes, ears, and nasal passages, or even can cause injury to the skin or mucous membranes of the mouth and nose simply by their mechanical presence. For this reason, ACGIH has recommended a limit of 10 milligrams per cubic meter for total dust containing less than 1 percent quartz for those substances for which no specific threshold limit has been allocated.

SIMPLE ASPHYXIANTS

As we have already seen, some gases or vapors, such as carbon dioxide, are not specifically toxic but can cause problems simply by blocking the available air supply. No PEL can be allocated for these substances because the limiting factor is not the toxicity but interference with the availability of oxygen. Oxygen normally occupies 21 percent of the air volume at sea level. To provide an adequate oxygen supply that would permit all the normal processes of living, it is necessary to ensure that the oxygen content of the atmosphere should be no less than 18 percent, a level equivalent to the conditions of breathing air at 8,000 to 10,000 feet.

DETERMINATION OF DAILY EXPOSURE

It is common practice to refer exposures to the level that would be expected over an eight-hour working day. As already noted, the threshold limit values refer to the levels permissible for an eight-hour day, five-day working week.

A detector tube, of course, provides a spot sample and, consequently, merely serves as an indication of the presence of a contaminant and its concentration at that particular time. A dosimeter badge, on the other hand, or a cassette with sample pump, when processed by a laboratory after being worn during a normal eight-hour day's work, can readily provide a measure of the extent of exposure of that worker during that period. It should be recognized, of course, that a dosimeter badge or cassette defines the overall accumulation. It gives no indication of whether during that period the worker was exposed at times to less or more than the average as calculated over the eight hours.

If the concentration during a working day is likely to vary considerably from the average, several shorter samples can be taken, either by using dosimeter badges or, when skilled professional help is available, by using a sampling pump and cassette. The results will then show the differences that might occur during the course of the day. A time-weighted average can be calculated from these sample results, by using the following simple formula:

$$E = \frac{(C_1 T_1) + (C_2 T_2) + \ldots + (C_n T_n)}{8}$$

where   E = 8-hour TWA exposure
   C_1 = concentration during the first sample period, assuming the concentration is constant
   C_2 = concentration during the second sample period, assuming the concentration is constant
   C_n = concentration during other and last sample periods
   T = time (hours) of each exposure
   8 = an 8-hour working day

Let us suppose that several air samples are collected containing the solvent methyl butyl ketone (MBK) over an eight-hour period with results as follows:

3 hours at 20 milligrams per cubic meter
2 hours at 16 milligrams per cubic meter
2 hours at 24 milligrams per cubic meter
1 hour at 0 milligrams per cubic meter (time off)

The resulting eight-hour TWA would then be:

$$E = \frac{(3 \times 20) + (2 \times 16) + (2 \times 24)}{8}$$

= 17.5 milligrams per cubic meter

If this number were then compared with the allowable exposure level of 20 millligrams per cubic meter (the derivation of which will be discussed later), you would know that the worker in question was not at risk.

Although the purists would argue that no exposure to a toxic chemical should be permitted (and in some cases this is justifiable), the realists will accept that under controlled conditions some minimal exposure is acceptable. The approach described above allows us to determine what that minimal exposure should be. At the same time, we must recognize that wherever possible, exposures, either in time or in concentration, should be reduced as close to zero as possible.

As previously noted, the PEL or TLV concept is widely used throughout the world in spite of some well-founded objections, largely because there is no feasible alternative. The most significant objection pertains to the use of the averaging technique. It is argued that the effects on the body are not necessarily the same when they result from constant uniform exposure to low concentrations as they would be when they result from intermittent exposure to higher concentrations over the same period. Nor is the dose received by various susceptible target organs in the body necessarily directly related to the average exposure level.

In the meantime, while management, labor, and scientists work to improve the system, this is the best we've got, and it's certainly a lot better than nothing.

CHAPTER SEVENTEEN

Step Five: Control the Exposure— The Administrative Approach

REQUIREMENT AND CONTROL

If you are aware of a health problem in your plant, you have to fix it, or at least control it, as soon as possible, for several reasons. One reason is self evident: it's the humanitarian thing to do. If conditions are known to be causing illness or death, then according to the mores of our society, these conditions must be improved. The other reasons are perhaps less obvious, but to some people more compelling. These reasons are to reduce costs and to ensure legal compliance. The question of costs has already been discussed. The true costs of occupational illness, other than those attributable to increased costs of workers' compensation, insurance dues, and possible penalties, are difficult to access but studies have indicated that the improved employee health and morale associated with improvement in healthy working conditions has ultimately led to improved productivity.

The question of legal compliance is another matter. Today, most developed countries and states have laws and regulations governing working conditions and the control of toxic materials in the workplace. While it may not be feasible to implement all of these requirements everywhere, at all times, the possibility always exists that even in a small plant to which nobody has paid much attention, the law may ultimately step in and demand costly changes and penalties that could have been avoided. The operative term under the law is "due diligence." Not only must the appropriate executives of a corporation, or the supervisors of a process, or an owner/proprietor ensure compliance with the letter of the law, they must also exert "due diligence" in ensuring the safe handling, storage, transportation, and disposal of dangerous and toxic materials. Courts have held that these obligations can extend beyond the immediate supervisors to individuals as distant as the directors of a company. Therefore, control is necessary, if only for legal reasons.

The process of control is one of applying the principles of industrial or occupational hygiene.The objective of industrial hygiene, as defined by the American Industrial Hygiene Association, is to ensure ". . . the recognition, evaluation, and control of those environmental factors and stresses, arising in or from the workplace, which may cause sickness, impaired health and well being, or significant discomfort and inefficiency among workers or among the citizens of the community." We have seen in previous chapters how to recognize and evaluate toxic conditions. Now we have to examine how to control them.

As was already noted in the introduction to Step One, Chapter 13, three approaches to control are available:

- administrative control
- engineering control
- personal protective equipment

In practice, some combination of all three is generally the most useful. It is a common and erroneous concept, however, to consider the primary approach to control to be the provision and use of personal protective equipment such as masks, gloves, and air supplies. This is just not so. In fact, although the use of personal protective equipment may well be desirable and even necessary, it should not be considered a primary approach except in short-term, temporary conditions.

ADMINISTRATIVE CONTROL

Ideally, one should eliminate a toxic problem or problems by administrative interdiction, or, in other words, by removing the cause. This is rarely possible. Commonly the process that gives rise to a problem is a necessary part of production. Nevertheless, many administrative actions can be undertaken which may not totally eliminate a problem but may go far to bring it under control. Some of these are broad actions that serve to establish a frame of reference within which other actions can be implemented. Others are more specific and are directed at control of an actual problem. The principles involved are as follows:

- Adopt a formal company policy of health awareness.
- Establish a management system for health related activities.
- Ensure that work practices are compatible with maintenance of health.
- Provide medical surveillance where necessary.
- Develop appropriate training programs.
- Control exposure of personnel to toxic conditions.
- Ensure good housekeeping and hygiene.

Each of these is considered below in detail.

ADOPT COMPANY POLICY OF HEALTH AWARENESS

Probably the first and most important administrative action is to adopt a company policy of health awareness. The attitude of top management to health awareness is reflected in the subsequent attitudes of middle management, supervisors, and the employees themselves. No program of control is going to succeed without the full cooperation of management. In order to establish the frame of reference within which further actions can be successful, it is necessary then to adopt, declare, promulgate, and publicize a formal policy of health awareness within the plant. This policy should include the following:

- maintenance of health of employees at work is fundamental to the continued activities of the company
- attainment and maintenance of a toxic-free environment is a primary company objective
- compliance with federal, state, or other regulations pertaining to the control of toxic materials is a company intent
- management, line supervision, support staff and employees are each and all responsible for ensuring healthy conditions and maintenance of a toxic-free environment

Once a policy has been established, it should be posted prominently and publicized regularly to all concerned by way of bulletin boards, meetings, and pamphlets, to ensure that all management and employees are aware of it. Not only must hourly employees be aware of it, but management must set an example by heeding its provisions, recognizing and acting on reports and recommendations, and motivating continued interest.

This type of organization, of course, cannot be developed without cost. Money may be required for personnel time that might otherwise be used for different purposes or even for the hire of professionals or technicians, as well as for the cost of equipment, materials, and additional training. It is important, as the program develops, that these and other related costs be recognized and provided with a specific budget.

APPOINT A HEALTH OFFICER

Depending on the size of the organization, it is desirable to designate an individual to act as full- or part-time health officer for the management of health-related activities. A large corporation might well have a corporate or local industrial hygiene department or industrial hygienist who could assume this responsibility. For a small company, however, this officer might be an owner, an owner/manager, a member of human resource staff, a safety officer, an occupational health nurse, or anyone with some health experience or someone who is able and willing to acquire such.

DEFINE HEALTH OFFICER'S DUTIES

The responsibilities of the appointed health officer are directed towards monitoring the environment and the employees working there to ensure the provision and maintenance of a healthy environment. They should include the following:

- advising in-plant departments such as engineering, line supervision, human resources, and purchasing about any potential hazard arising from an existing or proposed process
- recommending methods to eliminate or reduce employee exposure to harmful conditions
- establishing permissible levels of exposure (e.g., TLV's) and conducting periodic tests to ensure that these standards have been met
- monitoring existing and proposed practices to ensure they comply with regulations and established standards
- approving from a health standpoint all new processes and chemical usage
- assisting in the education of employees in healthy work practices
- specifying type, quality, and usage of personal protective equipment
- making available to all of those concerned the appropriate regulations, company policies and practices, MSDS's, and relevant documentation
- recording and investigating health-related accidents and incidents and making appropriate reports to higher management
- making regular reports to higher management. (If the health officer is an owner/manager, it is still advisable to file regular reports if only for future reference or legal reasons)

As with any new project, one of the first duties of the designated officer should be to conduct a careful study and analysis to determine the most appropriate type of program to be instituted, a study that begins with the walk-through survey already discussed. We are not so much concerned here with large plants of several hundred employees. These generally have a more or less effective system in place. For smaller plants, however, with no system in place, such an analysis is desirable. It should, for example, take into account the size of the organization, the nature of the hazards, and the availability of knowledgeable personnel. The health officer at this stage may find it necessary to seek some outside assistance. Such assistance can be made available from private consultants, safety organizations such as the National Safety Council, insurance companies that carry workers' compensation, the state industrial commission, or a health and safety council.

All of this cannot be achieved overnight. It can, however, be achieved over a period of time, but it will be apparent that the selected health officer should have appropriate training (although not necessarily professional) and should be influential enough to have the ear of those who can get things done.

For all situations that require the handling of toxic materials, the employees concerned should be provided with an instruction manual, prepared if necessary by a consultant, describing all foreseeable situations, routine and emergency, and the actions that should be taken if they should occur.

A trained and/or experienced occupational health nurse can be extremely valuable. Depending on circumstances and the size of the plant, a nurse could be either full-time or part-time. A full-time nurse is probably desirable for a plant employing 200 or more persons. Her duties, of course, would not be merely handing out medications and patching up injuries. She can undertake hazard identification and monitoring, medical surveillance and follow-up, occupational health incident investigation, workers' compensation and insurance liaison, consultation and advice with other management, union representatives, and exterior medical services, and can act as the designated health officer or provide valued assistance to a designated health officer. Out of this, she can provide reports to management and develop and maintain a record system of hazard existence and control as well as the health status of employees.

ENSURE THAT WORK PRACTICES ARE COMPATIBLE WITH HEALTH

To develop health awareness by instigating a formal policy for all levels is one thing; to ensure that such a policy is implemented is another.

To achieve this goal, top management must delegate authority, as necessary, to all management and supervisory levels. Supervisors in turn, as advised by the health officer, must accept and interpret the policy and actively support it. A supervisor, for example, must ensure that each of his or her employees understands the properties and hazards of any material that is handled, stored, or used by them, and takes all necessary precautions and follows established work procedures. An employee must make an effort to be aware of the hazards associated with toxic materials and follow the procedures laid down for their handling.

PROVIDE MEDICAL SURVEILLANCE WHERE NECESSARY

Some procedures and practices, for example the handling of lead, may require that employees undergo medical surveillance. This is never a practice that is well-accepted by employees, but one that nevertheless has to be implemented, sometimes under the penalties of law. This surveillance, of course, has to be done with the advice and participation of medical personnel. If the plant is large enough, and where practicable, much of this work can be undertaken by a trained occupational health nurse, whose skills may also be invaluable in other areas of health management. Otherwise, the requirement may have to be farmed out to an outside source.

DEVELOP TRAINING PROGRAMS

Before personnel at all levels can knowledgeably adopt healthy practices, they must be aware of what is required of them.

In addition to any normal training for their job, employees working with toxic chemicals need a formal training program or programs pertaining to the potential hazards and their management.

This training should include:

- a broad understanding of the nature and effects of toxic materials in general
- a specific understanding of the nature and effects of chemicals handled by the employees in their day-to-day work
- study and understanding of relevant MSDS's and other documents
- the significance of government regulations and the role of managerial and personal responsibility
- the practice of relevant administrative and engineering controls, and the use of personal protective equipment
- the need for and methodology of medical surveillance where appropriate
- the requirements of routine work practices and emergency procedures

CONTROL EXPOSURE OF PERSONNEL

It is essential to ensure that employees are not exposed to toxic materials at levels above or for durations beyond those permitted by TLV's or the equivalent. It is thus necessary to ensure by periodic measurement that atmospheric levels of toxic materials throughout the work shift do not exceed the permitted levels. Measurement of atmospheric toxic exposures has been discussed in Chapter 11. The frequency of this air sampling depends on the potential for exposure and injury. Where the hazard is relatively low and actions have been taken to control it, occasional checks by a knowledgeable but not necessarily professional person are desirable to ensure that the control is effective. Where there is a potential for serious hazard, for example, where high-hazard materials are handled in quantity, then continuous monitoring may be necessary. Control in the latter situation would normally demand the services of a skilled professional.

CONSIDER ROTATION OF EMPLOYEES

The question of isolation of toxic process will be considered when examining engineering approaches to control. However, there are situations where, despite all feasible control methods, exposures could exceed eight-hour TLV levels, or where the conditions are excessive but temporary. In these situations, when all else fails, consideration should be given to rotating employees through the task for such durations that no employee exceeds his/her short-term exposure level.

ENSURE GOOD HOUSEKEEPING AND HYGIENE

Much unnecessary exposure to harmful materials arises from the accumulation of dust, other solids, and liquids around a workstation, as well as open containers of solvents and other chemicals stacked in open storage areas. Enforcement of cleanliness and tidiness can do much to reduce this unnecessary exposure. A regular cleanup schedule should be instituted along with immediate cleanup of spills, covering of open containers, and general tidy practices. Toxic wastes should not be allowed to accumulate. Arrangements should be made for their proper disposal.

Handling of toxic materials may leave residues on the hands. Where the chemicals are irritant, skin reactions are inevitable. Perhaps even more significant, although much less obvious, when the hands touch the lips, in a casual gesture or during the process of coughing, sneezing, eating or smoking, this residue can be transferred to the lips and ultimately swallowed. Over a period of time, harmful materials can thereby accumulate internally to a toxic level. Where harmful materials are in use, it is essential that suitable washing facilities be provided and that workers be encouraged to wash their hands, especially before eating. Eating and, of course, smoking on the job, should be prohibited, and a suitable eating area should be made available, separate from the shop floor.

Administrative actions to control exposure to toxic materials are clearly vital, and no program of exposure control can succeed without them. They are, however, indirect, in that they do not directly affect the specific causative conditions. For direct action, we need to consider an engineering approach. This will be presented as a continuation of Step Five in the next chapter.

CHAPTER EIGHTEEN

Step Five: Control the Exposure— Engineering Control

GENERAL APPROACH

RESPONSIBILITIES

There are several ways in which engineering methods can be used to control exposure. In general, however, those persons responsible for engineering should ensure that all exposures are minimized to the extent feasible. In particular, existing processes are monitored and modified as far as reasonably possible to eliminate or reduce preventable exposure. The health officer or health and safety organization should be advised when a new process or use of a new and unfamiliar chemical is planned, and new installations should be checked by the health officer or organization before employees are permitted to operate them. For that matter, the design of new plants and processes should also be approved from a health point of view while still in the design stage to ensure, again to the extent feasible, that all systems and components are so designed as to contain all chemical contaminants within permissible limits.

MAINTENANCE

An important principle in process design is to ensure that periodically the equipment can be shut down for maintenance, process change, or process addition. This means that provision should be made for ensuring that the equipment is nonhazardous before it is dismantled or modified, by cleaning it with water, steam, or some neutralizing agent. This in turn requires that the equipment must be free of components or places where toxic materials might lodge in spite of routine cleaning, and also that it should be checked for the presence of toxic materials before any employee access is permitted for maintenance, repair, or modification.

ALARM SIGNALS

In some instances also, for example in the processing of isocyanates, which are highly toxic in minute quantities and generally processed inside an enclosure, it is desirable to consider the use of alarm signals to indicate the occurrence of a leakage or to demand the evacuation of an area.

Alarm signals can take the form of sounds or lights. Sounds can be sirens, buzzers, bells, and so on. Lights may be clear or colored. However, the selection of an alarm must take into account the ambient conditions. For example, a sound signal must significantly exceed the ambient noise level, and a light signal must be visible to all concerned under all lighting conditions.

Evacuation signals commonly take the form of a siren or buzzer, which after a ten second or more exposure may be followed by a voice message. The alarm and message should be repeated until the area is evacuated.

Signal lights are commonly used to direct attention to some immediate need. Color is relatively unimportant, although certain stereotypes exist such as red for danger. In point of fact, however, red is generally not as readily perceived as white or yellow. Location is important. If a light has to be seen by a given population, it should be within the visual range of the persons concerned, in other words, at eye level within about 30 degrees of the line of sight. The intensity should also be greater than the background intensity, and the signal should be clearly discriminable from the background. A large light is more readily perceived than a small light. Flashing lights are more attention-getting than steady lights, unless the background includes flashing lights. The optimum rate of flash is about 3 to 10 per second.

SPECIFIC METHODS

Besides ventilation, which will be considered later in this chapter, the actual engineering methods that may be used on the shop floor include the following:

- removal of a harmful material by substitution of a less harmful material
- change or modification of a harmful process or operation to minimize exposure
- isolation of a harmful process or operation to minimize exposure
- enclosure of a harmful process or operation to minimize exposure
- wetting down to reduce dust

Each of these is considered below. The purpose of this discussion, however, is not to outline in detail how to implement any particular approach. That would be impracticable. Each situation must be considered on its merits in the light of the potential harm intrinsic in the situation and the cost-effectiveness of a control measure. The intent here is to describe the principles involved so that potential approaches can be meaningfully considered.

SUBSTITUTION

Once a hazardous substance has been identified, it may be feasible, if suitable from a production viewpoint, to consider exchanging that substance for another less hazardous. This approach is often advantageous in dealing with solvents. Indeed, in some instances where materials are water soluble, simple detergent and water can be both a useful and less costly replacement. As another example, carbon tetrachloride, still in use in some places, can be replaced with methyl or ethyl chloroform or some other aliphatic hydrocarbon, as long as you remember that these too are toxic, although less so. Toluene can replace benzene with equal effectiveness in lacquers, synthetic rubber solutions, and paint removers. Acetone, which is relatively nontoxic in comparison with many other solvents can be an effective replacement in various instances. Various harmless powders can be used in foundries to replace silicon sand, while silica grinding wheels can be replaced, for instance, with artificial abrasive wheels containing aluminum oxide. Hand rubbing with mercury, too, is now almost never found in the felt hat industry.

CHANGE OF PROCESS

Most process changes are made for production reasons, but advantage can be taken of an opportunity during a production change to reduce any associated toxic exposure. If the existing condition is serious enough, a strong recommendation for change to take place at the earliest opportunity should be made even though it was not contemplated for production purposes.

As an example, grinding solder seams with high speed rotating sanding discs can generate lead dust. The occurrence of lead dust can be greatly reduced by changing to low-speed oscillating sanders. Similarly, where feasible, spray painting, which introduces large quantities of airborne mist into the atmosphere, can be changed to brush painting or dipping, or better still, electrostatic application if such is appropriate under the circumstances. Other examples include arc welding to replace riveting and ventilated vapor degreasing to replace open hand-washing of parts. Clearly, each situation has to be considered on its merits, and many other examples may be found.

ISOLATION

Some intrinsically toxic processes cannot be changed or modified. In these situations, isolation of the process can be considered. Isolation requires that the process or equipment be separated in some manner from nearby workers. The separation can take the form of an actual physical barrier that prevents an unprotected worker from approaching, or a distance barrier whereby the process can be operated by remote control, or even a time barrier whereby physical presence of the worker is only required for short periods during the working day.

Isolation has its greatest application in situations where relatively few workers are required or where more sophisticated forms of engineering control might be infeasible or not cost-effective.

ENCLOSURE

Enclosure takes the concept of isolation still further. In this approach, the complete operation is physically enclosed by an impervious barrier, perhaps metal or plastic, with access allowed only occasionally by suitably protected maintenance workers or in emergencies. This approach might be found, for example, in sand-blasting operations, in conditions where there is a severe fume hazard, or perhaps, as seen in the cautionary tale about Jake Whitman, in a continuous lead-quenching bath operation. It should be noted that in most instances, exhaust ventilation of the enclosed area should also be undertaken.

A modified form of enclosure exists in a paint booth where water is used as the physical barrier. In addition, however, the booth should be curtained off.

Next to substitution, isolation and enclosure are two of the most effective means of preventing or minimizing the escape of toxic vapor in the atmosphere, and should be the first measures used if substitution is infeasible.

WETTING DOWN

Dust can be stirred up by treading or other movement and can permeate the atmosphere and be inhaled. The dust that one can see can give rise to irritation and even dermatitis, but the dust that is inhaled can accumulate in the lungs and elsewhere and give rise to a variety of diseases from lead poisoning to metal fume fever.

A relatively simple method for removing dust accumulated on the floor is by wetting it down with water, perhaps aided by a wetting agent available from any supply house. The moist dust can then be swept up into a suitable container. Additional vacuum cleaning may be necessary when the dust accumulates in places other than the floor.

VENTILATION

Provision of adequate ventilation is one of the keys to development and maintenance of a contaminant-free workplace. Design of a proper ventilation system is normally a task undertaken by a professional engineer or a qualified industrial hygienist. All that will be attempted here is to present the principles so that the topic can be considered with some meaningful background.

The objective of ventilation is to ensure an adequate supply of fresh air at the workplace along with removal of airborne contaminants. There are two basic types of ventilation: *dilution ventilation,* or general ventilation, and *exhaust ventilation,* or local ventilation.

DILUTION VENTILATION

In dilution ventilation, quantities of clean air are provided to flood the airspace and reduce the contamination. The incoming air is allowed to flow across the relatively clean areas of the plant towards the contaminated areas from whence it is ultimately exhausted to the exterior.

In a properly designed and operating dilution system, the incoming air is at a higher pressure than the room air and consequently it will:

- expel and replace the local air
- dilute airborne dust, mists, and fumes
- prevent cross-contamination between defined areas by reason of pressure differences
- allow for heating, humidification, cooling, and draught reduction

Ideally the air will be introduced under controlled conditions of pressure and flow to provide for maximum dilution benefits. Because it is difficult, if not impossible, to measure or calculate the exhaust rate through doors, windows, leaks, and so on, it is desirable to allow for 10 percent more air to be supplied than the assumed exhaust.

The method of exhausting the air can be general or local. General exhaust may occur by way of gravity and natural pressure change through doors, windows, and so on, or it may be assisted mechanically at local sites by fans, commonly in the roof or side walls.

General dilution ventilation has the advantage of being relatively cheap to install and operate, but it is most effective only in the control of small quantities of mild to moderately toxic gases, mists, and fumes. It does not eliminate exposure, however, and it does not specifically control the exposure of the worker at the source. Furthermore, because of their relative density, it does not control heavy particulates or high density metal fumes, nor can it compensate for local changes in the generation of contaminants.

OTHER FACTORS TO CONSIDER IN DILUTION VENTILATION

PEL/TLV or equivalent. Where the PEL/TLV or equivalent is low, the potential for harm is greater and the usefulness of dilution ventilation is less. In this connection, a high TLV is considered to be 500 ppm and above, a moderate level 100 to 500 ppm, and a low level less than 100 ppm.

Rate of evaporation. Where the rate of evaporation is high, the potential harm is greater.

Location of worker with respect to the hazard. The closer the worker is to the operation, the less effective is the dilution ventilation.

Effectiveness of clearance. Where the general level of air flow at the workstation is poor, as in a corner, the less effective is the general ventilation.

Number of contaminants present. If more than one contaminant is present, the effects should be considered as additive.

EXHAUST VENTILATION

Exhaust ventilation is used to remove contaminants by suction at the source, transport them to a collecting area, and dispose of them. An exhaust system is more expensive to install and maintain than a dilution system, but because it is specific to the source or sources of contamination, it is more effective in its action.

An exhaust system has four basic components, namely, hoods, ducts, collectors, and fans. It may also incorporate an air cleaner.

Hoods

The function of a hood is to concentrate the suction as close to the source of contamination as possible. It should be placed as close to the work as feasible while ensuring that the worker's breathing zone is not contaminated, for example, by forcing him to place his head close to the hood or even beneath it.

To assist in concentrating the flow, the hood may have a flange, or broad flat rim, around the expansion of the hood bell, or it may have baffles or curtains to reduce unwanted air currents. While often placed at the back of the work area, drawing contaminants away from the worker, it may also be placed over the work area where the natural flow of air is upwards as in hot conditions.

The rate of air flow, or *capture velocity,* is normally calculated by the engineer or hygienist, but Table 18-1 gives an indication of the kind of capture velocity required.

Table 18-1
Range of Capture Velocities

Condition	Example	Capture Velocity (fpm)
Quiet air, very low velocity	Evaporation from tanks	50–100
Moderately still air, low velocity	Spray booths, intermittent container filling	100–200
Rapid air, active generation	Spray painting in shallow booths, barrel filling	200–500
Very rapid air, high initial velocity	Grinding, abrasive blasting, crushing	500–2000

Source: Developed from text in American Conference of Government Industrial Hygienists, Committee on Industrial Ventilation, Industrial Ventilation: A Manual of Recommended Practice. Cincinnati, OH, 1982.

The column labeled "condition" indicates the extent of air movement and an assessment of the rate of generation of contaminants. The column labeled "capture velocity" gives the rate of air flow at the hood required to capture the contaminants.

Where air currents are minimal, where contaminants are of low velocity or nuisance value, where the production of contaminants is intermittent or low, and where there is a large hood with a large mass of air in motion, then the lower range of capture velocity should be selected. On the other hand, where the air currents are disturbing, where the contaminants are generated at high velocity, where there is a high production of contaminants, and where there is a small hood for local control only, then the higher range should be chosen.

Ducts

Ducts are used to transport the exhaust air and its contaminants to the collection area. An important feature of duct design is reduction of resistance to air flow from friction within the wall, aggravated by unnecessary curves and bends (particularly right-angled bends), as well as junctions with other ducts at sharp angles, collections of dust, and changes in duct diameter.

While ducting is normally designed by an engineer or industrial hygienist, certain principles can be usefully considered here. Clearly, to reduce friction, the length should be as short as possible with the minimum of curves, bends, and junctions. Where angles are necessary, they should be no less than 135° if possible, and no junctions should be made at less than 45°. Because the caliber of ductwork at the beginning of the system, that is, the exhaust fan, is much larger than the caliber at the end of a branch, that is, a hood, the duct must change in diameter as it proceeds along the system. Any changes in duct diameter that are required as the duct approaches the source should be gradual rather than sudden to reduce obstruction and avoid collection of dust at the joining of two different calibers.

Leakages in the system should be prevented or eliminated. Leakages not only reduce the available suction but also allow contaminants to disperse into the surrounding area. To maintain the integrity of the system and reduce leakage and clogging, a regular schedule of maintenance should be instituted.

Collectors

A variety of different types of collectors may be used, depending on circumstances. Dry collectors, for example, are widely used in different types of plants and are suitable for conditions where the dust is relatively free from moisture. They can be installed and operated at medium cost and high efficiency. They are normally located outside the plant and are made up of a box with an inverted conical base or hopper. Inside the box are hung a number of tubular or envelope-shaped filters through which the air is passed. The filters may be made of cotton, wool, paper, glass cloth, or synthetics. Dust is built up on the filters until the resistance to flow becomes unacceptable, at which point the dust is extracted from the filters, commonly by vibration or reversal of flow, and allowed to drop into the conical hopper from whence it is manually or automatically removed.

A cyclone collector is another type of dry collector and is in the form of an inverted cone, but with no box. The incoming air containing contaminants is passed into the cone where it is spun at high speed. The spinning causes the solids to settle out to the periphery and fall into the apex of the cone for removal. Because of the relative densities, it is most effective with larger coarse particles, and not useful for fine particles.

There are several types of wet collectors including spray towers, packed towers, and wet centrifugal collectors. The spray tower is a cylindrical or rectangular tower into which the incoming air is passed. High-speed water sprays in the tower impact and remove the dust that is subsequently separated from the droplets by various types of eliminators. Spray towers are effective for all kinds of dust and even moisture-laden gases.

Packed towers are commonly cylindrical in shape and contain a series of plastic or ceramic beds or weirs over which the water sequentially flows. The air is passed across or through the beds. The system is particularly useful for gases, vapors, and mists, but less so for solid particles that tend to clog the flow.

A wet centrifugal collector is similar in principle to the dry centrifugal or cyclone collector except that the materials, gas, mist, vapor, or particulate is collected on wet surfaces.

Fans

The function of the fan is either to introduce broad general ventilation into an area or to pull air through a hood and ducting system to a collection device. There are two basic types of fans: the *axial* fan and the *centrifugal* fan. The axial fan is the type of fan that is commonly used by consumers. The air is drawn through the fan, past the fan blades, or impeller, and discharged. It is of greatest use in dilution ventilation and for cooling. It has relatively little directional flow because the discharged air, if not confined, tends to disperse widely within a very short distance from the blades.

In a centrifugal fan, the air is pulled into a housing along the line of the fan blades and then diverted to discharge. There are three types of centrifugal fans, depending on the angle of the blades to the incoming air.

The forward curved fan has blades that are curved forward in the direction of rotation. It is of best value in generating air flow for dilution ventilation and for exhaustion of air with no particulates. Particulates tend to accumulate on the blades, increasing resistance to a point where the fan is no longer effective. It has the advantage, however, of operating at relatively low speed, with consequently less noise, and relatively low cost.

The backward curved fan has blades that are curved backward against the direction of rotation. They are useful with light dusts, fumes, and condensates. They operate at higher speed than the forward curved fans and hence tend to be noisier and somewhat more costly, but, as a compensatory advantage, they are responsive to situations of varying resistance without loss of efficiency.

Radial fans, perhaps more descriptively known as paddle wheel fans, have blades located at right angles to the direction of flow. Because they do not clog as readily as the other two types, they are widely used in conditions where there is heavy dust or otherwise sticky materials. They are reasonably cost-effective, and because they operate at medium speed, they are also somewhat less noisy.

As mentioned previously, it is extremely important to recognize that personal protective equipment should not normally be considered the primary approach to protection against toxic materials. Nevertheless, such equipment is a vitally important mode. Its characteristics are considered in the next chapter.

CHAPTER NINETEEN

Step Five: Control the Exposure— Personal Protective Equipment

REQUIREMENTS

Personal protective equipment (PPE) is the term used for a variety of physical devices used to protect the body from hazards. Industrial hazards include impact, excess noise, heat, cold, and noxious chemicals of many different kinds and actions. The type of PPE we are concerned with here, of course, is equipment that can provide protection against hazardous chemicals. Protection may be required specifically for the face and eyes, skin, and the respiratory system, and each may need a different kind of safeguard. However, as previously stated, except in emergencies and special circumstances, PPE should not be relied on as the primary or sole approach to protection. Primary attack at the source by administrative and engineering means must always be considered paramount.

SELECTION OF EQUIPMENT

Respiratory protection is protection against gases, fumes, mists, vapors, and particulates that can enter the respiratory tract through the mouth and nose and either settle in the air passages and lung or be taken into the blood stream and transmitted through the body.

According to Section 1910.134(b) of the Occupational Safety and Health Standards in Title 29 of the U.S. Code of Federal Regulations, an acceptable respiratory program must be instigated whenever respiratory health hazards are present in the work environment. The employer must develop formal written operational procedures covering every aspect of the program, including, for example, how contaminants are controlled, how contaminant concentration is measured, and how respirators are selected, used, cleaned, inspected, repaired, and stored.

Choosing the proper respirator is essential. Some of the criteria are discussed below. Help for making the selection can be found if necessary by way of private consultants, insurance companies, NIOSH and OSHA personnel, and even from respirator manufacturers and distributors.

It should be recognized that protective equipment does not eliminate the hazard, it merely prevents contact with it. If the device is imperfect or inadequate or becomes so while in use, the user may be unexpectedly exposed to conditions where escape is not immediately possible.

The authoritative source of information for respiratory protection is the American National Standards Institute, *Practices for Respiratory Protection,* ANSI Z88.2-1992 (see Appendix II for address).

In making a selection, consideration should be given to the factors described below.

FACTORS TO CONSIDER WHEN SELECTING A RESPIRATOR

Nature of the hazardous materials. Consideration of the nature of hazardous materials includes recognition of:

- their physical properties
- their effects on the body
- their concentration in the atmosphere
- their limits of permissible exposure

These characteristics have been examined in previous chapters. Now is the time, however, to reconsider them in order to make a proper choice of protective equipment. As we shall see when we come to look at the characteristics of the equipment, it would be dangerous to select a dust filter when a chemical cartridge respirator is needed, or a chemical cartridge respirator when a self-contained unit is needed. Several other considerations, such as those that follow, should be examined.

Expected duration of exposure. The requirements for short duration exposure can be different from those for long duration. A task, for example, that requires 30 minutes of work in an enclosed space with toxic gas might well be satisfied by a cartridge respirator. For prolonged exposure, a mask with an attached air supply hose and blower might be more appropriate.

Location of the hazard area. Air supply for a hose mask or air line respirator requires a source of clean, uncontaminated air within the distance limits of the air hose, that is, 300 feet for a hose with an attached air blower and 75 feet without a blower. Obviously, before selecting a respirator of that type, the source of clean air should be identified.

Functional characteristics of the devices. These will be examined later.

Functional characteristics of the employees. The wearing of respiratory devices, except for simple dust masks, can be both physically and psychologically demanding. Even a cartridge respirator without any requirement for a hose connection can be restrictive, the more so in the presence of any obstruction to breathing by a clogged filter. Indeed, to some susceptible persons, the very presence of a face mask can be claustrophobic. In addition, the need to drag around 200 to 300 feet of hose or a heavy self-contained apparatus similar to scuba divers' equipment can be physically demanding. Persons with respiratory and/or cardiac problems such as asthma, bronchitis, angina, or with other debilitating conditions should not be permitted to work in conditions where respiratory equipment is required. The emotional status of workers should also be checked to the extent feasible, and only volunteers used in severe conditions.

You should also note that respirators cannot be properly used if the seal between face and facepiece is broken by sideburns, eyeglass side pieces, and even scars and wrinkles and missing dentures.

Support and maintenance. Facilities must be available for the proper storage and maintenance of respiratory equipment. Regular cleansing and sanitation is essential. It is preferable that each employee be assigned his/her own respirator, but, failing that, the user must be satisfied that the respirator is in fact clean and sanitary. All equipment, with the exception of consumable dust filters, which should be replaced after use, must be kept clean and ready for use. Over a period of time, a piece of equipment will deteriorate even when not used. Consequently, regular inspection must be conducted; damaged or deteriorated equipment must be identified and replaced when necessary. It is also necessary to ensure a supply of consumables such as replaceable filters as well as equipment components.

Storage should be provided in clean bags or other suitable containers within a clean and sanitary location.

Monitoring of Program. A regular procedure of monitoring, with respect to changing conditions, usage, cleanliness, and storage should be instituted to ensure that the respirators are appropriate to the conditions and are being properly used and properly maintained.

Training in care and use. Respiratory equipment can be misused by ignorance, negligence, or default. Consequently, training in the care and use of equipment is essential for all the employees concerned. This may have to include training in the use of emergency equipment for workers who might not otherwise be involved. Training may be conducted by knowledgeable persons within the plant, or by safety organizations, either government or private, or even under the auspices of the manufacturers.

TYPES OF RESPIRATORY EQUIPMENT

As indicated above, the type of respiratory protective equipment to be used depends, among other factors, on the nature of the hazard.

Respiratory protective equipment can be divided into two broad categories: air purifying and air supplying devices. Air purifying devices are items such as dust masks, chemical cartridge respirators, and gas masks. The air supplying devices are hose masks, air line respirators, and self-contained apparatus.

AIR PURIFYING DEVICES

Dust masks

Dust masks, or particulate filter respirators, protect the respiratory system against any kind of inhaled particles. The simplest type, in the form of a cone-shaped fibrous filter that fits over the nose and mouth, is held in place by an elastic band around the back of the head. It is disposable, relatively comfortable to wear, and is useful when the dust is neither excessive nor toxic. It becomes readily clogged and should be discarded and replaced whenever breathing becomes restricted. Its use should not be carried over from day to day.

More sophisticated versions are cartridge masks that have rubber or synthetic facepieces that cover the mouth and nose and are attached to the head by straps. Flat cylindrical cartridges containing dust filters are attached to each side of the facepieces. The cartridges come apart for easy removal and replacement of the filters. The devices are useful in heavy, continued dust conditions. The cartridge filters should be replaced whenever they become clogged.

Air Purifying Respirators

Air purifying respirators are similar in style to the more sophisticated dust masks, and may incorporate dust protection. They have a properly fitting facepiece with a removable colored flat cylinder that acts as both filter and chemical neutralizer attached to each side. Sometimes only one cartridge is mounted in a central position. The facepiece may be a half facepiece covering only the mouth and nose or a full facepiece covering the full face and eyes as well, depending on individual requirements. Cylinders are interchangeable and vary in size and color codings according to the protection requirements. As many as 20 different types of cartridges are available for different purposes. Commonly used cartridges are listed in Table 19-1.

Certain limitations should be noted. First, respirators do not provide protection when the oxygen supply is reduced, nor should they be used in

Table 19-1
Commonly Used Respirator Cartridges

Color	Purpose
Black	Organic vapors
Black with white and gray stripe	Acid gas: chlorine, hydrogen chloride, sulfur dioxide, hydrogen sulfide, formaldehyde
Orange	Organic vapors, chlorine, hydrogen chloride, sulfur dioxide
Green	Ammonia/methylamine
Brown	Formaldehyde

With an added white prefilter these cartridges can also be used in the presence of dusts and mists with a time-weighted average (TLV-TWA) of not less than 2 million particles per cubic foot.
Source: Developed from National Standards Institute ANSI Z88.2-1922.

situations where the concentration of the contaminant (according to NIOSH approval limits) is greater than ten times the permissible exposure or the maximum use concentration prescribed in applicable government standards, whichever is the lower. Second, a cartridge designed for protection against one contaminant will be effective for longer than the same size canister designed for protection against more than one contaminant. Third, the higher the concentration of the contaminant and the greater the activity of the user, the shorter will be the length of protection provided.

It is important to recognize that cartridge respirators should be used only for protection against those contaminants listed on the filters or cartridges or on their cartons. In particulars, it is important to realize that cartridge respirators do not protect against carbon monoxide, which is a common and dangerous industrial contaminant.

Cartridge respirators are also not suitable for use in firefighting and sandblasting as well as in atmospheres containing less than the normal 21 percent of oxygen or where the concentration of the contaminants is unknown or immediately threatening to life. In these unsuitable situations, an air line respirator or a self-contained apparatus (see following) should be used.

Masks can be used on repeated occasions provided the cartridge is not exhausted. A cartridge approaching exhaustion can be identified in use by odor, taste, or eye, nose, and throat irritation while in use. Wherever odor or irritation is encountered when wearing a cartridge respirator or where the wearer feels dizzy or breathing becomes difficult, the wearer should immediately leave the area and replace the respirator.

Gas masks

A gas mask commonly comprises a rubber or synthetic facepiece connected by a short flexible tube to a canister that lies on the chest. The facepiece is carefully fitted to the face over the mouth and nose and sometimes the eyes and is held in place by straps around the head. The canister is supported by shoulder and chest straps. In another version, a smaller and lighter canister is suspended from the facepiece harness and hangs directly below it. Despite its name, the gas mask is used not only in protection against gases, but also against fumes and other vapors, depending on the nature of the active chemical within the canister. Canisters are color coded in the same manner as cartridges. Because canisters are larger than cartridges, they provide a longer protection than do the cartridges.

AIR SUPPLYING DEVICES

Air supplying devices are not designed to purify the breathing air. Instead, they supply breathable air either through a hose leading to a suitable air supply or by a self-contained apparatus. When properly fitted and free of inward leakage, they can be used in the presence of toxic gases, mists, and other vapors in concentrations above those described for air purifying respirators or in an oxygen-deficient atmosphere.

Hose respirators

The hose respirator is the least expensive of the air suppliers, but is of limited usefulness. It comprises a carefully fitted full facepiece and one or two flexible corrugated rubber tubes that lead from the mask to the accompanying hose. The hose may be a breathing tube, not more than 75 feet long, open to a source of clean air. A hose longer than 75 feet requires more suction than inhalation can supply and creates what is called a large *dead space.* More commonly, the breathing tube may be equipped with a hand-operated or power-driven blower supplying air through up to 300 feet of hose. Even with a blower, however, the hose respirator has an uncomfortable, limiting, and fatiguing restriction to breathing that can be both physically and psychologically demanding. The hose is pulled by way of a body harness.

The device is of particular use for entering tanks or pits where dangerous contaminants or deficient oxygen may be suspected. The location, of course, should be tested for contaminants and oxygen before a worker enters it and ventilated if necessary. Care should also be taken to ensure that the user is physically and emotionally fit to use the equipment in hazardous conditions.

Air Line Respirators

Air line respirators are similar to hose respirators, except that they are supplied with compressed air from a clean air source and compressor. They provide complete protection, and because they offer little in the way of breathing resistance, they are more suitable than hose respirators for prolonged use. There are two basic types: continuous flow and demand flow.

In the *continuous flow* version, a regulated amount of air under low pressure is fed continuously to the facepiece. In the *demand* version, the air passes through a regulator that governs the pressure and flow. From the regulator, the incoming air is inhaled through an inhalation valve that opens on initiation of a breath by the user. The expired air is discharged through an exhalation valve. A pressure relief valve is also incorporated into the system to relieve pressure in the event of regulator failure. Normally the facepiece is kept under slight positive pressure to prevent inward leakage from the contaminated atmosphere.

Although still awkward and slightly claustrophobic to wear and physically demanding because of the need to drag a hose, the system provides complete protection in all atmospheres. Other than a self-contained apparatus, this system is the most desirable for conditions requiring prolonged use of a respirator.

For conditions where the atmosphere is also corrosive to the skin and mucous membranes, for example in the presence of hydrochloric or hydrofluoric vapors, a ventilated impervious suit incorporating a breathing air supply should be used. Ventilation is required for cooling.

Self-contained Apparatus

A self-contained breathing apparatus (SCBA) should be used in atmospheres that are immediately hazardous to life. It should be carefully maintained and stored when not in use in accordance with the instructions of manufacturers or safety and health institutions and government directives. Care should be taken to ensure that the wearer is physically and mentally fit, well trained, and aware of potential hazards.

There are two main classes of SCBA: the closed circuit, or recirculating, and the open circuit. The former is more expensive to purchase but less expensive to operate, particularly if using oxygen. There are two types of the closed circuit class: the oxygen generating type and the compressed gas (oxygen or air) type.

Closed Circuit Oxygen Generating

This device comprises a properly fitting full facepiece with tube assembly leading into a breathing bag or reservoir that is filled as required from an attached chemical canister in which the oxygen is generated. Flow is regulated by check valves. A pressure relief valve is also incorporated. The apparatus is simple and relatively light. The action of the canister is initiated by the user when the chemical in the canister comes in contact with the moisture and carbon dioxide from the exhaled breath. The chemical continues to generate oxygen and absorbs the moisture and carbon dioxide.

Commonly, more oxygen is generated than required and it has to be periodically vented. The volume is generally good for up to an hour. A used canister should be discarded after use and a new canister fitted.

Closed Circuit Compressed Gas

The compressed gas may be air or liquid oxygen. The equipment is composed of a facepiece, regulator and valves, a breathing bag, a scrubber for removing carbon dioxide from the exhaled air, and a portable gas cylinder worn on the back. The purified exhaled air is rebreathed from the reservoir bag and replenished to the extent necessary from the supply tank. The equipment can be used in any atmosphere, but is complicated to operate and maintain. Thorough training is required both of the user and the person responsible for maintenance. Consequently, while perhaps the best type of equipment for continued use, it is expensive. The cost of operation is reduced, however, because the recirculating system reduces the requirements for breathing gas.

Open Circuit Apparatus

The open circuit apparatus is essentially similar to the closed circuit compressed gas system, except that the user exhales through an exhalation valve to the exterior rather than into a recirculating system. Because the exhaled air contains unused oxygen as well as unwanted carbon dioxide, the open circuit, while initially cheaper, is less efficient than the closed circuit.

OTHER PROTECTIVE EQUIPMENT

Safety equipment such as helmets, steel capped boots, and hearing protectors are, of course, important and should be used when the situation warrants. Our discussion here, however, is concerned with the protection of the eyes and skin in the presence of chemicals that are irritant, corrosive, or capable of penetration.

FACE AND EYE PROTECTION

Total protection from liquids and vapors can be provided by an impervious hood incorporating a glass or plastic window. Unfortunately, these are very hot and may require additional ventilation directly from an air line, or perhaps through a device known as a vortex tube that converts supplied compressed air to either a cool or warm flow as required.

When less protection is required, it can be provided by a face shield either attached to a hard hat or held in position by a head harness that allows the shield to be tipped back when not in use. Flexible fitting chemical goggles with hooded ventilation can be used for direct protection of the eyes.

GENERAL SKIN PROTECTION

Ordinary clothing ultimately becomes permeated or sometimes even destroyed by chemicals, some much more readily than others. If the clothing is destroyed, as perhaps by strong acid or alkali, the chemicals concerned will have direct access to the skin. Otherwise, the permeated clothing can either irritate and ultimately injure the skin or allow penetration of the skin by the chemicals involved if they have that capacity. Consequently, in the presence of dusts, vapors, and corrosive liquids, some form of impervious clothing is desirable, in the form of aprons, bibs, or whole body garments. Materials used include natural and synthetic rubber, vinyl, polypropylene, and polyethylene, either alone, in combination with natural and synthetic fabrics, or as coating films. It should be noted that natural rubber deteriorates in the presence of oils, greases, and many organic solvents.

HAND PROTECTION

The skin of the hand comes in contact with hazardous chemicals more frequently than any other part of the body and is more in need of protection. Normal work gloves are inadequate protection against hazardous chemicals. Materials for protection should be the same as those noted above. Gloves should be long enough to come well above the wrists and leave no gaps between glove and arm coverings. They should fit snugly about the wrist.

Protective creams are commonly used to provide a barrier between the skin and potential contaminants. While effective for short periods where the material is of low toxicity or penetrability, or where there is minimal skin contact, or where gloves are otherwise impracticable, they are not very satisfactory for prolonged use because the barrier, if it is effective at all, tends to be effaced with continued exposure. They are very useful as an adjunct lubricant before donning gloves and act as a protection should the gloves be damaged in use.

It should be remembered, of course, that ordinary soap and water can provide a very good deterrent to skin irritation and injury. Only the very strong chemicals cause immediate damage to the skin. Frequent and thorough washing can go far to reduce the possibilities of skin damage.

Hazards from Various Processes

The following are not inclusive, but represent typical potential hazards.

Table A-1
Hazards from Various Processes

Process	Potential Hazard
Abrasive blasting	Dust
Babbitting	Fumes, dusts: antimony, tin, lead
Bagging, handling, dry	Dusts
Ceramic coating	Dispersion of pigments (cadmium, chromates, lead)
Coal handling	Dusts, gases: carbon monoxide, coal, silica, sulfur dioxide
Coke handling	Dust: coke, silica
Coking	Multiple: ammonia, benzene, carbon disulfide, carbon monoxide, cyanamides, hydrogen

(continued on next page)

Table A-1 (Continued)
Hazards from Various Processes

Process	Potential Hazard
	sulfide, naphthalene, phenols, pyridine, sulfur dioxide
Dry grinding, mixing	Dusts
Fabric, paper coating	Solvents
Forming and forging	Lubricant mists and decomposition products
Galvanizing	Fumes, gases, mists: ammonium chloride, chromates, hydrochloric acid, hydrogen chloride, lead oxides, zinc oxide
Molten metals	Gases, metal fumes, dusts
Paint spraying	Solvents, pigments (cadmium, chromates, lead)
Pickling	Gases, mists: hydrochloric acid, hydrogen chloride, hydrogen fluoride, oxides of nitrogen, sulfuric acid
Plating	Mists, gases, direct contact: acids, alkalis
Refractories handling	Dust: silica
Sintering	Dusts, gases: carbon monoxide, fluorides, free silica
Solvent degreasing	Vapors: perchloroethylene, trichlorethylene, vapor decomposition (e.g., phosgene)
Steelmaking	Dusts, fumes: lead oxide, iron oxide, fluorospar, graphite, limestone, ore
Welding	Fumes, gases: oxides of cadmium, chromium, fluorides, iron, manganese, nickel, nitrogen, vanadium, by-products from fluxes, coatings, electrodes

Source: Derived in part from U.S. Department of Labor: Occupational Safety and Health Administration, Washington, D.C., Compliance Operations Manual, *OSHA-2006, 1972.*

Short-Term Detector Tubes

The following tubes are a selection of those used to measure the instantaneous concentration of gases and vapors in the atmosphere. Tubes of this type are manufactured and distributed by the Dräger organization in Europe (Draeger in the U.S. and Canada), and by the National Mine Service Company in the U.S. They can be obtained from a safety supply store.

Acetaldehyde
Acetic acid
Acetone
Acrylonitrile
Alcohol
Ammonia
Aniline
Arsine
Benzene
Carbon dioxide
Carbon monoxide
Carbon tetrachloride
Chlorine
Chlorobenzene
Formic acid
Formaldehyde

Hexane
Chloroformates
Chloroprene
Chromic acid
Cyanide
Cyanogen chloride
Cyclohexane
Diethyl ether
Dimethyl formaldehyde
Dimethyl sulfate
Dimethyl sulfide
Epichlorohydrin
Ethyl acetate
Ethyl benzene
Ethylene glycol
Fluorine

Olefine

Organic arsenic compounds

Organic nitrogen compounds

Halogenated hydrocarbons

Hydrazine

Hydrocarbons

Hydrochloric acid

Hydrocyanic acid

Hydrogen

Hydrogen fluoride

Hydrogen peroxide

Hydrogen sulfide

Mercaptan

Mercury

Methanol

Methyl acrylate

Methyl bromide

Methylene chloride

Nickel

Nickel tetracarbonyl

Nitric acid

Nitrous fumes

Oxygen

Ozone

Pentane

Perchloroethylene

Petroleum hydrocarbons

Phosphoric acid esters

Phenol

Phosgene

Petroleum hydrocarbons

Pyridine

Styrene

Sulfur dioxide

Tetrahydrothiophene

Thioether

Toluene

Toluene diisocyanate

o-toluidine

Trichlorethane

Triethylamine

Vinyl chloride

Water vapor

Xylene

Useful Sources

PUBLICATIONS

American Conference of Government Industrial Hygienists, P.O. Box 1937, Cincinnati, OH 45211-4438

Threshold Limit Values

Guide to Health Records for Health Services in Small Industries

Industrial Ventilation—A Manual of Recommended Practice

Air Sampling Instruments for Evaluation of Atmospheric Contaminants

American Industrial Hygiene Association, 66 South Miller Road, Akron, OH 44311-1087

American Industrial Hygiene Association Journal Hygienic Guides (to control many different chemicals)

Respiratory Protective Devices Manual

American Medical Association, Circulation Department, 535 N. Dearborn Street, Chicago, IL 60610

Archives of Environmental Health

Guide to Small Plant Occupational Health Programs

American National Standards Institute, 1430 Broadway, New York, N.Y. 10018

Identification of Air-purifying Respirator Canisters and Cartridges, K13.1

Practice for Occupational and Educational and Face Protection, Z87.1

Practices for Respiratory Protection, Z88.2

Manufacturing Chemists' Association, 1825 Connecticut Avenue, N.W. Washington, DC 20009

Chemical Safety Data Sheets

National Safety Council, 425 N. Michigan Avenue, Chicago, IL 60611

Industrial Skin Diseases

Industrial Data Sheets (Summaries of toxic data and control strategies on many industrial chemicals)

Pergamon Press, Inc., Maxwell House, Fairview Park, Elmsford NY 10523

Annals of Occupational Hygiene

U.S. Department of Health Education and Welfare, Public Health Service Center for Disease Control, National Institute for Occupational Safety and Health, Morgantown, WV 26505

Guidelines on Approved or Certified Personal Protective Devices and Industrial Hazard Measuring Instruments

U.S. Department of Health Education and Welfare, National Institute for Occupational Safety and Health, Cincinnati, OH 45226

NIOSH/OSHA Pocket Guide to Chemical Hazards, 1990

U.S. Government Printing Office, Superintendent of Documents, Washington, DC 20402

Toxicity Bibliography

BOOKS OF INTEREST

Anon. *Accident Prevention Manual for Industrial Operations.* Chicago: National Safety Council, 1974.

Anderson, K. and Scott, R. *Fundamentals of Industrial Toxicology.* Ann Arbor, MI: Ann Arbor Science Publishers Inc./The Butterworth Group, 1981.

Fraser, T. M. *The Worker at Work.* London, New York, Philadelphia: Taylor and Francis, 1989.

Goldman, F. H. and Jacobs, M. B. *Chemical Methods in Industrial Hygiene.* New York: Interscience Publishers, 1953.

Proctor, N. H. and Hughes, J. P. *Chemical Hazards of the Workplace.* Philadelphia: J. P. Lippincott Company, 1978.

ORGANIZATIONS, AGENCIES AND SOCIETIES

American National Standards Institute (ANSI), 1430 Broadway, New York, N.Y. 10018

ANSI is a federation of over 1,000 national trade, technical, professional, labor, and consumer organizations, governmental agencies, and individual companies. It coordinates the standards development efforts of these groups and implements national standards when they are approved. Among other publications, it publishes documents on health and safety.

American Conference of Government Industrial Hygienists (ACGIH) P.O. Box 1937, Cincinnati, OH

ACGIH is a professional association derived from industrial hygiene personnel in government at all levels or personnel working under a government grant. It has numerous committees dealing with different aspects of industrial hygiene, including some with the American Industrial Hygiene Association. It publishes numerous documents on topics in industrial hygiene, including, in particular, the *Threshold Limit Values.*

American Industrial Hygiene Association (AIHA), 66 S. Miller Road, Akron, OH 44313

AIHA is a professional association of industrial hygiene personnel with the objective of disseminating industrial hygiene knowledge in the field and promoting the study and control of environmental factors affecting the health of industrial workers. AIHA publishes *Hygienic Guides,* which summarizes current information on specific chemicals, as well as a professional journal and various reports and monographs.

Government Agencies

Departments or boards of health and industrial hygiene services are integral parts of the organization of all states, the District of Columbia, and the autonomous territories of the United States, as well as the federal and provincial governments of Canada and all developed nations.

Industrial Health Foundation, Inc., 5231 Centre Avenue, Pittsburgh PA 15232

The Industrial Health Foundation, with headquarters at the Mellon Institute, is a nonprofit research association of industries with the objective of advocating industrial health programs, improving working conditions, and encouraging better human relations. It provides direct professional assistance to member companies in control of industrial health hazards, assists companies in the development of health programs, and contributes to the advancement of industrial hygiene and medicine.

International Labour Office, (ILO) CH1211, Geneva, 22, Switzerland

Sometimes erroneously called the International Labour Organization, the office is an agency of the United Nations formed to improve labor conditions, raise living standards, and promote economic and social stability.

It has a specific mandate to provide "protection of the worker against sickness, disease, and injury arising out of his employment." It publishes the prestigious *Encyclopedia of Occupational Health and Safety,* as well as numerous monographs relating to occupational health. The Health Information Centre (CIS) analyzes and provides abstracts of relevant articles appearing in official publications and professional journals throughout the world.

Manufacturing Chemists' Association, 1825 Connecticut Avenue, Washington DC 20009

This association is concerned with all aspects of chemical manufacturing, but, in particular, provides service in disseminating information on the safe handling, transportation, and use of chemicals. It publishes *Chemical Safety Data Sheets* covering the hazards.

Index

Index